Integrated Reservoir Studies

Integrated Reservoir Studies

Editor

Monika Pandey

Integrated Reservoir Studies
Edited by **Monika Pandey**

Printed in 2017

ISBN: 978-1-68117-407-5

Library of Congress Control Number: 2015941613

© 2016 by
SCITUS Academics LLC,
616, Corporate Way, Suite 2, 4766,
Valley Cottage, NY 10989

www.scitusacademics.com

Contents

Preface

Integrated Reservoir Studies is to acquaint petroleum geologists and engineers with techniques used to characterize hydrocarbon reservoirs, with specific focus on the construction of static and dynamic reservoir models. The overall objectives will be to improve production rates, ultimate recovery, and field economics by identifying potential for infill drilling, recognizing bypassed hydrocarbons, and increasing recovery efficiency. When applied to reservoir studies, the integration process can help foster more effective, less expensive projects. What, however, does integration really mean, under everyday working conditions? Which new technical and professional challenges must be faced? What kind of changes does it imply for traditional ways of working? What is the role of the project leader in the integration process? The goal of this book is to provide some answers to these questions, and to highlight the critical differences between an integrated reservoir study and traditional one.

Editor

Chapter 1

A Practical Biodegradation Scale for Use in Reservoir Geochemical Studies of Biodegraded Oils

Steve Larter[a, b], Haiping Huang[b], Jennifer Adams[b], Barry Bennett[a], and Lloyd R. Snowdon[a, b]

[a]Gushor Inc., Bay #2, 925 – 30th Street NE, Calgary, Canada T2A 5L7

[b]Petroleum Reservoir Group, Department of Geoscience, University of Calgary, Canada

ABSTRACT

Existing scales widely used to describe the extent of biodegradation of petroleum have insufficient resolution to usefully characterize many heavy oil and bitumen occurrences, including the volumetrically dominant heavily and severely biodegraded oil accumulations in the foreland basins of western Canada and Venezuela. In these and other deposits, existing classifications or descriptions of the biodegradation level may vary only slightly, yet oil may vary in viscosity by orders of magnitude. The "Manco" biodegradation scale proposed here is based on integrating the extent of degradation of various members of compound classes not included in previous biodegradation scales. They include alkyl aromatic and alkyl thiophenic compounds that show variable extent of alteration in samples degraded to uniform levels on standard scales, but which may show variation in local degradation systematics related to biodegradation mechanisms and extent of oil mixing. The Manco scale uses a combination of a consideration of the extent of alteration within a compound class together with a consideration of biodegradation across a range of compound classes. It can be reliably used as a basis for interpreting geochemical changes in heavily biodegraded oil suites and can also be used to differentiate biodegraded oil samples likely to be amenable to cold production from those requiring production strategies such as steam or chemical flooding. As with other biodegradation scales, the scale may also provide evidence for the influx of later, higher quality oil into a reservoir fluid that had been previously biodegraded.

INTRODUCTION

Heavy oils, super heavy oils and oil sand bitumen are formed by microbial degradation of conventional crude oils over geological timescales. The final distribution of oil API gravity and fluid properties, such as the viscosity of oils found in heavy oil fields, is ultimately controlled by a complex interplay of oil charge rate, pervasive oil charge mixing, microbial diversity and oil chemistry due to compound specific biodegradation rates (controlled by temperature, water chemistry and nutrient supply; Head et al., 2003, Larter et al., 2003, Larter et al., 2006a and Adams et al., 2006). While biodegradation is a primary control on fluid properties in many heavy oil settings such as in western Canada, the nature of the primary oil charge and addition of a secondary fresh oil charge may dominate properties in settings such as in China and western Canada (Koopmans et al., 2002). All these processes complicate existing simple rankings of crude oils in terms of relative level of degradation.

Previous Biodegradation Scales

Peters and Moldowan (1993) developed a quasi-systematic classification scheme (PM scale; Fig. 1) to rank the level of biodegradation of an oil on a scale of 1 (least altered) to 10 (most altered). The PM scale has been widely and successfully used, especially for light, conventional oils and condensates. A gas chromatography (GC) trace [or gas chromatography–mass spectrometry (GC–MS) total ion current (TIC) chromatogram] of a non-degraded, conventional crude oil is visually dominated by n-alkanes, their presence indicating a pristine or slightly biodegraded oil. Alteration of diasteranes and high molecular weight (MW) aromatic steroid hydrocarbon compounds only becomes apparent when the oil has been very heavily degraded, i.e. at PM 9–10. Another scheme (Wenger et al., 2002) is based on a more comprehensive list of compounds, compound classes and carbon number ranges (Fig. 2), but uses only two categories ("heavy" and "severe" degradation) to describe all stages within the PM 4–10 range. These scales are essentially based on the presence or absence of single key compound classes and a consideration of the extent of alteration within a compound class is accommodated by using dotted or lightly shaded areas in the presence/absence graphs (Fig. 1).

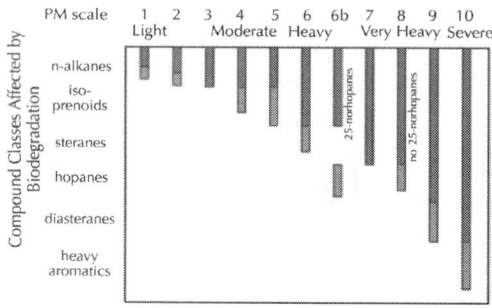

Figure 1: Biodegradation scale modified after Peters and Moldowan (1993). The lightly shaded areas at the ends of the bars attempt to reflect qualitatively the extent of partial removal of compounds within the class.

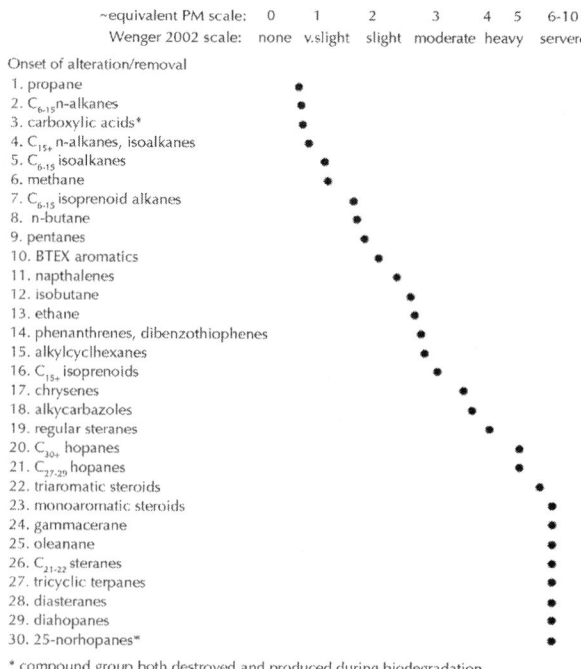

Figure 2: Removal of selected compound groups at various levels of bio-degradation [Head et al., 2003; modified after Wenger et al. (2002)].

Both these scales present problems in adequately describing the extent of biodegradation in super heavy oils (e.g. western Canada) because essentially all the heavy oils and Athabasca Tar Sand bitumens fall within PM level 5–8, that is, within the two Wenger categories of "heavy" to "severe" alteration. Peace River oil sand oils are almost all classified at PM level 5, but display orders of magnitude variation in their "dead oil" viscosity values, a property used to evaluate recovery strategies and carry out resource assessments. Knowledge of the extent of degradation is a critical parameter in the western Canada heavy oil belt because the choice of different exploitation strategies that may be applicable depends on the physical properties of the oil (largely viscosity), which correlate strongly with the extent of biodegradation. Applications are not merely practical ones, as biodegradation scales also allow geochemists carrying out more basic study to communicate the relative differences in the extent of the biodegradation process among related samples.

While existing schemes for biodegradation assessment (Volkman et al., 1983, Peters and Moldowan, 1993, Wenger et al., 2002 and Peters et al., 2005) have proven useful for conventional light oil, they were generally based on small calibration sets, often from different source facies and basins. Some of the calibration oils had undergone surface evaporation and extensive water washing (Peters and Moldowan, 1993), which preferentially alter an oil in a manner different from biodegradation. They also did not explicitly consider the dominant role of mixing of differently biodegraded oils in the genesis of final fluid composition, nor did they address local variability in biodegradation systematics such as the possibility of aerobic vs. anaerobic degradation (Wardlaw et al., 2011) or more likely, different anaerobic degradation pathways (Head et al., 2003). Additionally they did not address the range of variability in oil quality observed at the higher levels of biodegradation common in the volumetrically important oil sands and heavy oil belts of western Canada, Venezuela and elsewhere. In western Canada the oil deposits are commonly biodegraded to a level of PM 5–8 but have highly variable physical properties (e.g. dead oil viscosity varies from thousands to tens of millions centipoises at 20 °C) even though the PM level shows little variation.

Summary of Current Views on Oil Biodegradation

Large scale lateral and vertical variation in biodegraded oil properties within reservoirs are common, due to the interaction of biodegradation and charge mixing (Horstad and Larter, 1997, Larter et al., 2003, Larter et al., 2006a, Larter et al., 2006b and Larter et al., 2008a).

Generation is the rate limiting step in reservoir charging and source rocks typically charge petroleum traps over an interval of the order of a few million to a few tens of millions of years (Pepper, 1991). The notion of a single oil charge to any oilfield is, however, illusory insofar as it could be argued that the oil in any reservoir is an integrated product of various fractions of early, middle and late generation/migration of oil arriving typically over timescales of a million years or longer (Larter et al., 2003). While it is sometimes convenient to consider that the charge arrives episodically, in reality charging is a continuous, if sometimes interrupted, process during source rock expulsion. Thus, by definition, all petroleum accumulations are comprised of mixtures of oils, even before any biodegradation occurs.

Petroleum biodegradation produces compositional variation within oilfields and within individual petroleum columns. Biodegradation proceeds at the oil–water contact (OWC) under anaerobic conditions in any reservoir that has a water leg and has not been heated to >80 °C (Head et al., 2003) and proceeds on a similar timescale to oil charging (Larter et al., 2003). Net degradation of petroleum fractions in reservoirs is primarily controlled by reservoir temperature, the structures of the compounds being degraded and relationship between the OWC surface area and oil volume. Relative volumes of bottom water to oil leg, reservoir water salinity and prior levels of oil biodegradation act as second order controls (Larter et al., 2006b). Typically, degradation flux for fresh petroleum in clastic reservoirs is in the range of 10^{-5}–10^{-3} kg petroleum degraded/m^2 OWC area/yr and rises with decreasing reservoir temperature, from zero near 80 °C, to a maximum flux of $<10^{-3}$ kg petroleum/m^2 OWC/yr at <40 °C (Larter et al., 2003) Nutrient supply from the reservoir aquifer and adjacent shales, mostly buffered by mineral dissolution, probably provides a major control on the range of values of degradation flux.

Biodegradation of all compound classes occurs simultaneously but at different rates (Larter et al., 2006b). That is, the absolute concentration of all of the susceptible compound classes is reduced simultaneously as biodegradation proceeds. However, the *relative* concentration of more resistant compounds and compound classes will increase by virtue of the very rapid loss of mass associated with more labile fractions. The net result of progressive biodegradation *appears* to be a sequential removal of different classes of compounds rather than actual simultaneous removal on different rate trajectories. For light crude oil, the normal removal sequence appears to start with low MW *n*-alkanes followed by higher MW *n*-alkanes, followed in turn by branched alkanes such as the acyclic isoprenoids and then more resistant compounds such as cyclic alkanes. The extent of relative removal generally shows a broad correlation with the physical properties of the oil (e.g. viscosity; Koopmans et al., 2002).

Variation in oil quality controlled dominantly by biodegradation may significantly impact on exploitation strategy. For instance, subsurface biodegradation in a reservoir can produce up to a 50× increase in dead oil viscosity over a distance as short as 40 m within a single reservoir (Larter et al., 2008a), affecting well placement, and higher viscosity oils with dead oil viscosity of many thousands of cP (at 20 °C) may require the use of thermal stimulation using steam or electrical heating to allow production.

Adding fresh oil to an existing biodegraded oil significantly changes key compositional parameters (e.g. relative concentration of biomarkers) but also has a significant impact on the physical properties of the oil mixture. Later oil charges are typically more mature and relatively enriched in gas and light ends relative to the initial charge, so the oil in a subsequent charge effectively acts as a solvent. A relatively small amount of "solvent" (e.g. 5 wt% toluene) added to an oil of 100,000 cP dead oil viscosity will cause a significant reduction in the viscosity of the mixture (typically an order of magnitude) but only a modest increase in API gravity (Larter et al., 2009 and Jiang et al., 2010). Changes in the physical properties may be predicted if charge mixing can be quantified accurately or, conversely, the charge history may be determined from a careful consideration of the spatial distribution of the oil properties. The processes and their impact on oil composition and fluid properties have been summarized by Head et al. (2003) and Larter et al., 2006a and Larter et al., 2006b.

The distribution of oil columns subjected to biodegradation is such that basal oil, near the OWC, is usually the most degraded and has the lowest API gravity and the highest polar compound content and viscosity, while the best quality oil occurs near the top of the reservoir (Larter et al., 2008a). Samples from a single oil column in a single well may show vertical variation in apparent biodegradation level that can be dramatic and require careful consideration of the resolution of scale used to describe the extent of biodegradation. Much of the variation in the composition of samples in the middle of an oil column represents vertical mass transport of components downward towards the OWC (Larter et al., 2003), where they are removed through biodegradation. That is, the vertical transport is largely driven by chemical potential (concentration gradients) established by way of the removal of selected compounds through biologically mediated reactions operating at the bottom of the oil leg, near the OWC.

Additionally, or alternatively, compositional gradient may also reflect differential or competing rates of reservoir filling and simultaneous biodegradation during the filling of the reservoir, with commensurate downward or upward migration of the OWC as the reservoir fills, depending on the relative rates of biodegradation and reservoir charging (Larter et al., 2003, Larter et al., 2006b, Larter et al., 2008a and Larter et al., 2008b). The introduction of a second (or subsequent) charge of fresher oil may also be locally constrained (i.e. within only selected compartments in a reservoir) and have an impact on the apparent local extent of biodegradation.

In the Athabasca oil sands, biodegradation has completely removed the n-alkanes, branched alkanes and 1–2 ring aromatic hydrocarbons and can affect even the cyclic alkane biomarkers (Bennett et al., 2005). These highly viscous oils are usually rich in asphaltenes and resins (commonly 40 wt% or more). Alkyl aromatic hydrocarbon degradation is one of the best systems for tracking bitumen quality variation in the oil sands. The changes correspond to highly viscous crude oils where dead oil [2] viscosity may exceed 10^6 cP at 20 °C. In the Peace River area, the presence of alkyl naphthalenes and alkyl toluenes indicates less extensive degradation and that the oils may be amenable to cold production with unaltered alkyl naphthalene distributions observed up to viscosity values of ca. 20,000 cP. Extensive alkyl naphthalene alteration and some alkyl phenanthrene alteration indicate that these oils will most likely need thermal recovery (steam heating) and will not be producible via cold or primary production. Significant alteration of alkyl phenanthrenes and significant increase in the 9-/1-methylphenanthrene abundance ratio suggests that the viscosity of the oil is likely to be well above the cut off for economic cold production in Peace River. In both Athabasca and Peace River oil sand provinces, the PM levels of oils within a single reservoir commonly vary little or not at all, despite the changes in fluid properties.

Many factors, including initial oil composition and oil mixing, may affect the final oil viscosity in a complex manner. Small amounts of light end compounds can disproportionately affect oil viscosity and, in many legacy oil viscosity data sets, loss of light ends from oils during sample storage can easily introduce errors of up to an order of magnitude in viscosity (Adams et al., 2008). It is this sample integrity issue that has prevented effective viscosity-compositional correlation unless specialized oil extraction procedures are applied (Larter et al., 2006b) and sophisticated multivariate methods are utilized for geochemistry-viscosity correlation (Adams et al.,

2010). Nevertheless, geochemical indications, even in the absence of calibration oils, can be used to provide general fluid property assessments, so biodegradation level assessment is commonly a routine part of heavy oil reservoir geochemical assessments in applied studies.

Why do We need a New Biodegradation Scale?

In general, existing biodegradation scales have insufficient resolution to allow useful characterization of the extent of biodegradation in many heavy oil fields, especially the western Canada heavy oil and bituminous sands belt and the so-called underlying "carbonate triangle" (Creaney et al., 1994 and other papers in the WCSB Atlas). This lack of resolution is limited by the classes of compounds used in the definitions of existing scales. Early scales provided a useful but broad, general level of degradation description across many oil types and had a predominant focus on saturated hydrocarbons, primarily in conventional light oils and condensates. For practical and more fundamental applications, improved resolution was important. For example, in the Peace River oil sands of Alberta, analysis of several hundred samples indicates that, while the biodegradation level ranges from PM 3–6, >90% of the samples would be described by a degradation level of PM 5, even within a single oil column and even though the dead oil viscosity at 20 °C ranges from a few thousand cP to >6 million cP (Adams et al., 2008). In response to the limitations imposed by existing biodegradation scales, we have developed a higher resolution scale ranging approximately through PM levels 4–8. The "Manco" scale is useful for describing the extent of biodegradation of heavy oil and oil sand bitumen in the Western Canada Sedimentary Basin and unpublished results suggest that it will probably be applicable to other heavy oil and super heavy oil deposits globally. The approach can also be generalized to other oil settings by way of the use of any sequence of progressively less or more biodegradation refractory compound classes to refine the resolution of existing models.

MANCO SCALE OF BIODEGRADATION

In the western Canada oil sands belt with biodegradation level ≥PM 5, it is typically the concentration and distribution of 1–3 ring aromatic hydrocarbons that most strongly co-vary with the physical properties of the oils. Oil production and upgrading or processing properties are also strongly covariant with these compound classes at intermediate levels of biodegradation, such as at Peace River (Bennett and Larter, 2008). On the other hand, for somewhat shallower reservoirs further to the east at Athabasca, aromatic hydrocarbons with three or more rings and cyclic alkane (naphthenic) biomarkers are more effective for tracking the biodegradation process, oil viscosity and other property variations.

To resolve the limitations of published scales, we have conceived a novel approach using integrated scoring of the biodegradation of several compound classes. The method is different from previous methods in three ways: (i) eight compound classes of differing reactivity, that are appropriate to heavy oil and bitumen, have been identified and used; (ii) the extent of alteration within a compound class has been used in at least a semi-quantitative manner and (iii) the extent of alteration within a compound class and between compound classes has been combined in an algorithm using a linear function within a compound class and a power function between classes. The total hydrocarbon fraction (non-polar fraction) from an oil sample (Bennett and Larter, 2000) was analyzed using selected ion monitoring (SIM) GC–MS for 60 ions at ca. 1.6 Hz with an Agilent 6890N GC instrument coupled to a 5975B mass spectrometry (MS) instrument. ChemStation software was used to control the instrument and to process the data for a range of saturated and aromatic hydrocarbons. Mass chromatograms of various compound classes were inspected and scores ranging from 0 through 4 (*Manco score: Modular Analysis and Numerical Classification of Oils*) were assigned (see below) for each of a suite of eight classes (Table 1). The distributions and scores within the classes change systematically with biodegradation in the range seen for heavy oil and bitumen samples from western Canada (*Manco vector*). A more general classification scheme could use any number of fractions characterized in any manner ordered in terms of generally increasingly refractory behavior. The Manco vector is then algorithmically summed to give a single *Manco number*, which provides an overall estimate of the extent of biodegradation of heavy oil or sand samples. The *Manco scale* is the range of Manco numbers (0–1000) defined for heavy oil and bituminous sand deposits generally relating to PM level 4–8. The maximum value of 1000 is arbitrary but serves to allow resolution of subtly different levels of biodegradation and provide absolute values outside of the range of existing scales, thus avoiding any confusion.

Table 1: Compounds and classes used as categories reflecting increasing resistance to biodegradation; *m/z* values of the ions used for GC–MS detection are also shown

Vector element	Compound class	GC–MS *m/z* value
0	Alkyl toluenes	105
1	C_{0-1} naphthalenes (N + MN)	128 + 142
2	C_2 naphthalenes (C2N)	156
3	C_3 naphthalenes (C3N)	170
4	Methyl dibenzothiophenes (MDBT)	198
5	C_4 naphthalenes (C4N)	184
6	C_{0-2} phenanthrenes (C0–2P)	178 + 192 + 206
7	Steranes	217

The approach ranks the level of alteration of different compound classes from key mass chromatograms by way of simple visual assessment scored into five levels from 0–4. For the Manco scale, the eight compound classes consist of 1-, 2- and 3-ring aromatic hydrocarbons (i.e. alkyl toluenes, naphthalene + methyl naphthalenes, C_2-, C_3-, C_4-naphthalenes, C_{0-2} phenanthrenes), methyl dibenzothiophenes and steranes. A Manco score of 0–4 may be readily distinguished qualitatively for any compound class from gas chromatograms, as completely non-degraded (Manco score 0) to fully degraded (typically the compounds are absent; Manco score 4), with classes not quite fully degraded (3), or only very slightly degraded (1) and oils mid-way between the extremes (2). This affords five levels of increasing degree of degradation (0–4) for each compound class described as 'pristine', 'light', 'moderate', 'heavy' and 'depleted' in terms of biodegradation for each compound group. Changes in these compound classes have been empirically observed to be applicable to heavy oils, extra heavy oils and bitumen samples collected from the Peace River area and also for much of the Athabasca oil sands as well as samples from other basins. Fig. 3 shows examples of five GC–MS traces for C_{0-2} alkyl phenanthrenes (m/z 178 + 192 + 206 mass chromatograms) with Manco scores from 0–4. Similar descriptions and scores are applied to all eight compound categories in Table 1. The scores are then assembled into an eight element vector (group of numbers) to which is applied an algorithm to calculate a single Manco number used to describe the overall level of biodegradation. A base 5 scale for assessing biodegradation level was chosen by polling several geochemists as to a common, universally agreed generic level of discrimination possible in visual assessment of gas chromatograms of degraded oils. Thus complete removal of a compound class (score 4 – depleted) or no alteration (score 0 – pristine) are easily agreed and it seems mild alteration (score 1 – light) and minor remaining presence of target compounds (score 3 – heavy) are also easily agreed. An intermediate (score 2 – intermediate) represents a mid point in the scale. Attempts at alternate score bases either failed to discriminate the variability seen (bases < 5) or exaggerated visual discrimination possibilities (base > 5). We believe a base 5 system is an optimal and pragmatic scale for visual biodegradation level assessment within a compound class and crucially is equivalent to the n-alkane removal scale used in the PM scale (i.e. 0, intact n-alkanes; 4, n-alkanes gone).

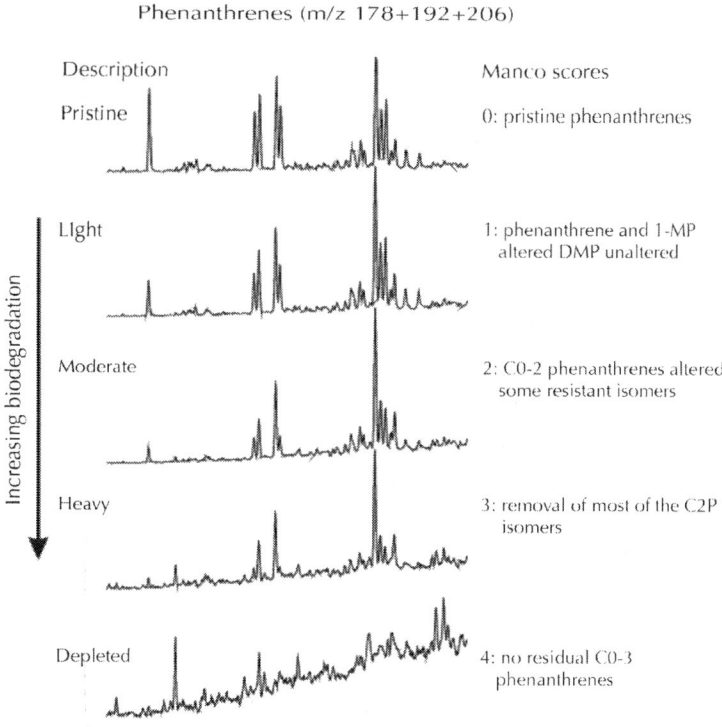

Figure 3: Examples of GC–MS chromatograms and associated Manco scores for different levels of biodegradation illustrated by the alkyl phenanthrenes.

To illustrate the approach, Fig. 4 shows the aromatic hydrocarbon (C_3 alkyl naphthalenes and phenanthrene) concentration in petroleum extracted from reservoir cores, which shows a downwards progressive increase in biodegradation (decreasing compound concentration) to the site of the OWC at ca. 675 m in a well from the Peace River oil sand province (note that the phenanthrene concentration is multiplied by a factor of 10 for visualization purposes). Dead oil viscosity from oils recovered from core samples is also shown for comparison, as are the Manco number 2 (MN2) values (see below for explanation) for the oils and an indication of the PM level which is uniformly 5 throughout the reservoir interval (for visualization purposes this is plotted as PM X

100). There is a general correlation between MN, the hydrocarbon concentration and the oil viscosity. The approach provides added resolution for monitoring the extent of biodegradation, particularly for extensively degraded oils. We note that, although sterane mass chromatograms do not change significantly through the oil column, there is a suggestion that sterane concentration does decrease somewhat at the base of the oil column, indicating fairly homogeneous degradation of steranes. However, the PM scale relates solely to chromatogram-based distribution assessments, so we have maintained the degradation level as PM 5. Below, we discuss the integration of compound concentration and distribution data together in biodegradation assessment and recommend that this is clearly the way forward, but today few labs routinely measure and use compound concentration data. Therefore, we have continued to use visually assessed compound distribution assessments as the primary, rapid biodegradation assessment tool here because that technology is widely available and routinely used.

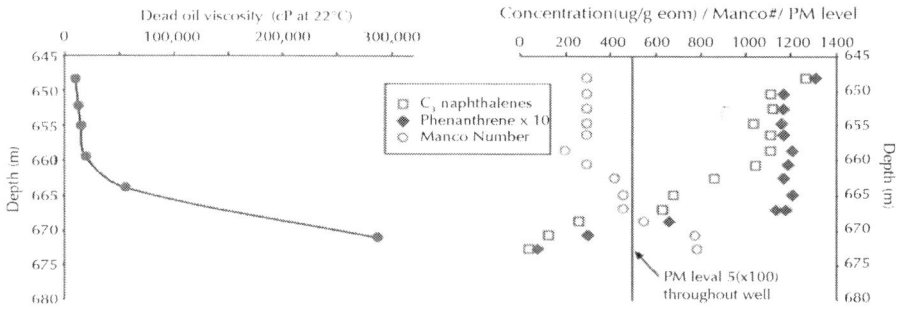

Figure 4: Aromatic hydrocarbon (C_3 alkyl naphthalenes and phenanthrene) concentration in petroleum extracted from reservoir cores shows a progressive increase in biodegradation extent down to the site of the oil–water contact (OWC) at ca. 675 m in a well from the Peace River oil sand province (note that the phenanthrene concentration is multiplied by a factor of 10 for visualization purposes only). Hydrocarbons diffuse towards the OWC, where they are metabolized by microorganisms using nutrients derived from the water saturated zone below the oil column. Fresh oil is charged to the top of the reservoir, while degradation occurs at

the base of the oil column. Compositional gradients reflect this complex charge and degradation scenario, with reactive compounds (e.g. reactive aromatic hydrocarbons decreasing in concentration towards the OWC). Measured viscosity from recovered oil recovered by spinning core samples in a high speed centrifuge is also shown, as are the Manco number 2 (MN2) values and an indication of the PM biodegradation level, which is uniformly 5 (for visualization this is plotted as PM 5 × 100). Note the response of the Manco number and hydrocarbon concentration to biodegradation and the general correlation with oil viscosity.

Table 2 shows the eight categories included in the Manco vector for 71 samples, or plausible theoretical cases, at different levels of biodegradation. The more refractory compound classes are given progressively more weighting in the calculation of the Manco number by observing that the Manco vector (the numerical aggregation of the compound class scores) is calculated as a base 5 number, so a MN value is obtained by multiplying the Manco score (0–4) for each category by 5 raised to the power of the appropriate category number (0–7) for compound classes of increasingly refractory nature.

As described above, experiments were carried out with other bases and the apparent optimum balance between the progression of alteration within a compound class (scores) and the progression through the different compound classes was achieved with a base of 5. The use of a simple linear function combining the scores (relative extent of alteration within a compound class) and the observation of any alteration within a compound class (essentially the way the current scales operate) cannot easily accommodate the fact that there is overlap in the alteration of adjacent compound classes. For example, the C_3 naphthalenes are partially altered before the C_2 naphthalenes are completely removed. The sequence of apparent alteration may be somewhat variable because of variable microbial communities, variable oil or water chemistry or a range of other factors including nutrient availability and temperature.

Table 2: Manco scores for 71 samples or theoretical cases showing 8 compound categories (i.e. the Manco vector), calculated Manco number and viscosity data (where available). In general, compound classes more susceptible to biodegradation would be expected to be more altered than more resistant classes (i.e. scores decreasing left to right in the vector). Where this is not the case (e.g. cases 8–12, 14, 16, 17, etc.), this may indicate a second charge of fresh oil into a reservoir containing biodegraded oil and/or unusual behavior of the microbiological consortium

#	Alkyl-tol	N + MN	C2N	C3N	MDBT	C4-N	C0-2 P	Sterane	MN2	Viscosity cP @ 20 °C
1	1	0	0	0	0	0	0	0	1	
2	1	1	0	0	0	0	0	0	140	
3	2	2	0	0	0	0	0	0	194	37600
4	2	2	0	0	0	0	0	0	194	45000
5	2	2	0	0	0	0	0	0	194	65000
6	2	2	0	0	0	0	0	0	194	46500
7	2	2	1	0	0	0	0	0	281	
8	2	3	1	0	0	0	0	0	291	32300
9	2	3	1	0	0	0	0	0	291	33000
10	2	3	1	0	0	0	0	0	291	35000
11	2	3	1	0	0	0	0	0	291	77000
12	2	3	1	0	0	0	0	0	291	
13	3	3	1	0	0	0	0	0	293	184000
14	3	4	2	0	0	0	0	0	334	75000

	335	0	0	0	0	0	2	4	4
48000	398	0	0	0	1	1	1	3	2
43000	409	0	0	0	1	1	2	3	2
73000	409	0	0	0	1	1	2	3	3
74500	409	0	0	0	1	1	2	3	3
90000	409	0	0	0	1	1	2	3	3
103000	409	0	0	0	1	1	2	3	3
216000	409	0	0	0	1	1	2	3	3
	411	0	0	0	1	1	2	4	3
	412	0	0	0	1	1	2	4	4
112000	419	0	0	0	1	3	3	3	3
166000	421	0	0	0	1	3	3	4	4
	455	0	0	1	2	3	3	4	4
758000	483	0	0	1	3	4	4	4	4
	521	0	1	1	1	2	3	3	3
93600	524	0	1	1	1	3	4	3	3
94000	524	0	1	1	1	3	4	4	4
115000	524	0	1	1	1	3	4	4	4
109000	524	0	1	1	1	3	4	4	4
129000	524	0	1	1	1	3	4	4	4
160000	534	0	1	2	2	3	3	3	3
	537	0	1	2	2	4	4	4	4
	575	0	2	2	2	4	4	4	4

Row labels (rightmost column): 15, 16, 17, 18, 19, 20, 21, 22, 23, 24, 25, 26, 27, 28, 29, 30, 31, 32, 33, 34, 35, 36, 37

38	4	4	4	2	3	0	0	0	600	
39	4	4	4	3	3	0	0	0	604	99000
40	4	4	4	4	3	0	0	0	608	136000
41	4	4	4	3	1	1	0	0	649	1E + 06
42	4	4	4	4	1	1	0	0	651	
43	4	4	4	4	3	1	0	0	671	
44	4	4	4	2	0	0	1	0	752	291000
45	4	4	4	2	0	0	1	0	752	396000
46	3	3	3	2	1	0	1	0	755	89600
47	3	3	3	2	1	0	1	0	755	84700
48	4	4	3	2	1	0	1	0	755	137000
49	3	3	4	2	1	0	1	0	755	119000
50	4	3	3	2	2	0	1	0	758	
51	4	4	4	2	2	0	1	0	758	
52	4	4	4	3	2	0	1	0	758	201000
53	4	4	4	3	2	0	1	0	758	197000
54	4	4	4	3	2	0	1	0	758	298000
55	4	4	4	3	2	0	1	0	758	315000
56	4	4	4	3	3	0	1	0	761	327000
57	4	4	4	4	4	0	1	0	764	275000
58	4	4	4	3	2	1	1	0	771	1E + 06
59	4	4	4	4	2	1	1	0	772	
60	4	4	4	3	3	1	1	0	774	

61	4	4	4	4	3	1	1	0	774	170000
62	4	4	4	4	3	1	1	0	774	244000
63	4	4	4	4	3	1	1	0	774	527000
64	4	4	4	3	4	2	1	0	786	628000
65	4	4	4	4	3	1	2	0	817	
66	4	4	4	4	4	2	2	0	824	
67	4	4	4	4	4	3	2	0	830	924000
68	4	4	4	4	3	2	1	1	896	2E + 06
69	4	4	4	4	4	2	2	1	908	4E + 06
70	4	4	4	4	4	3	3	2	954	
71	4	4	4	4	4	4	4	3	983	

The sum of the compound group scores is termed the MN 1 [Eq. (1)]. The number of compound classes plus log (base 5) of MN1 × the scale maximum minus one (999 in this case), is then divided by the number of compound classes (in this case 8, i.e. 0–7) to yield MN values ranging from 0–1000. This second number (MN2), is the biodegradation level descriptor [Eq. (2)]. Introducing the "number of compound classes" at two points in the calculation returns a value of 1 if MN1 = 1 and makes the approach general such that the numbers of classes of compounds may be varied (see Section 5) to accommodate different circumstances or different compound class combinations.

Heavy Degradation Manco Numbers

Manco number 1 : $MN1 = \Sigma(\text{Class score}_i \times 5^i)$
for compound classes $i = 0–7$ $\hspace{2cm}$ (1)

The compound classes are listed in Table 1. Scores of the different classes are multiplied by 5^i where i is the class number (0–7). The use of the power function reflects the fact that increasing alteration within a compound class (scores increasing from 0 through 4) is the result of relatively minor differences in extent of degradation, whereas the alteration of an increasingly resistant compound class reflects a very significant and non-linear increment in degradation. Previous biodegradation scales have simply applied a more or less linear category ranking based essentially on the presence or absence of a compound class. The calculation of MN1 for case #2 (base 5 Manco vector = 11000000; written as a base 5 number this would be 00000011) in Table 2 is $MN1 = (1 \times 5^0) + (1 \times 5^1) + (0 \times 5^2) + \cdots = 6$.

Manco number 2 : MN2
$\hspace{1cm}$ = [(number of compound classes)
$\hspace{1.5cm}$ + $(\log_5(MN1) \times (\text{scale maximum}$
$\hspace{1.5cm}$ − 1)]/(number of compound classes) $\hspace{1cm}$ (2)

MN1 = 0 is a special case where MN2 is assigned a value of 0 to accommodate $\log_5 (0)$, which is undefined. The number of compound classes is 8 and the arbitrary choice of a maximum MN2 of 1000 was made to avoid confusion between this and existing scales and to allow sufficient resolution at different levels of biodegradation using integer values. The $\log_5$ function is applied to the calculation to return the power function to a more or less linear overall scale while allowing for enhanced weighting of the degradation of more resistant classes of compounds. The spreadsheet formula for evaluating MN2 is thus: MN2 = IF{MN1 > 0 then [8 + (999 * log[MN1, 5])] else 0}/8.

The calculation of MN2 for case #2 (MN1 = 6) is thus MN2 = [8 + 999 * $\log_5(6)$]/8 = 140.

Case #26 in Table 2 is an oil with a viscosity of 166,000 cP (20 °C) with a Manco vector of 44310000 (written as a base 5 number this would be 00001344), which indicates that the alkyl toluene and naphthalene + methyl naphthalene classes are completely removed (scores of 4) while the dimethyl naphthalenes are severely altered (score of 3) but are not entirely absent and the trimethyl naphthalenes are only slightly altered (score of 1). The methyl dibenzothiophenes and more resistant compound classes have all been assigned a Manco score of 0, indicating that no alteration is apparent in the chromatograms. Manco number 2 is calculated as 421.

The Manco number is strongly controlled by the extent of alteration of the most resistant compound category that is affected by biodegradation. For example, samples #50 and #51 in Table 2 both have calculated Manco numbers of 758 despite the fact that for sample #50 the N + MN and C_2 naphthalene categories have lower scores (3) than interpreted values (4) for sample #51. Similarly, sample #52 has a Manco number of 758 despite having a higher Manco score for the C_3 naphthalenes than #51. Clearly, the Manco score of '1' for the C0–2 phenanthrenes dominates the calculated Manco number for all three samples. Thus the Manco number provides information on the most biodegraded phase of a mixed sample, as is the case for other biodegradation scales. Mixing or other unusual processes are brought to light by investigating the details of the less resistant compound categories within the Manco vector (Table 2; see below). Combining absolute component concentration data and the Manco numbers permits more detailed investigation of oil mixing phenomena as discussed below. The MN2 number has proved to be helpful in mapping bitumen quality for oil sand recovery in western Canada and for detecting changes in biodegradation style between areas.

The Manco scale effectively provides a means of discussing biodegradation levels within PM 5–8, or even just within the large range in oil composition found at PM 5 where much alkyl aromatic hydrocarbon and alkyl thioaromatic degradation takes place. In addition to the calculated Manco number for a number of examples with different Manco vectors, Table 2 also includes examples of viscosity values measured at 20 °C (extrapolated to 20 °C for samples #64 and #67). Fig. 5 shows a cross plot of $\log_{10}$ (viscosity) vs. MN2. Typically oils with a viscosity <100,000 cP have a Manco number 2 (MN2) of <500. The low viscosity oil samples with a Manco number of 755 (#46 and #47) are not depleted in alkyl toluenes or C_2naphthalenes but have suffered degradation of the C_{0-2} phenanthrenes, suggesting that these samples may have had later input of petroleum containing alkyl benzenes that effectively reduced the viscosity of an earlier biodegraded and more viscous oil. These samples plot below the general trend between Manco number and viscosity (Fig. 5). Sample #28 has an unusually high viscosity (758,000 cP @ 20 °C) for its Manco number of 483. This character is associated with a very sharp transition between almost complete degradation of the C_3 naphthalenes (Manco score 3) and pristine methyl dibenzothiophenes (Manco score 0). Other samples with anomalously high oil viscosity values also show a sharp transition in the level of degradation with increasing compound category number (Fig. 5). There is no obvious process or genetic cause that is apparent for these observations and this may be an artefact of the process or may indicate a key observation of process behavior.

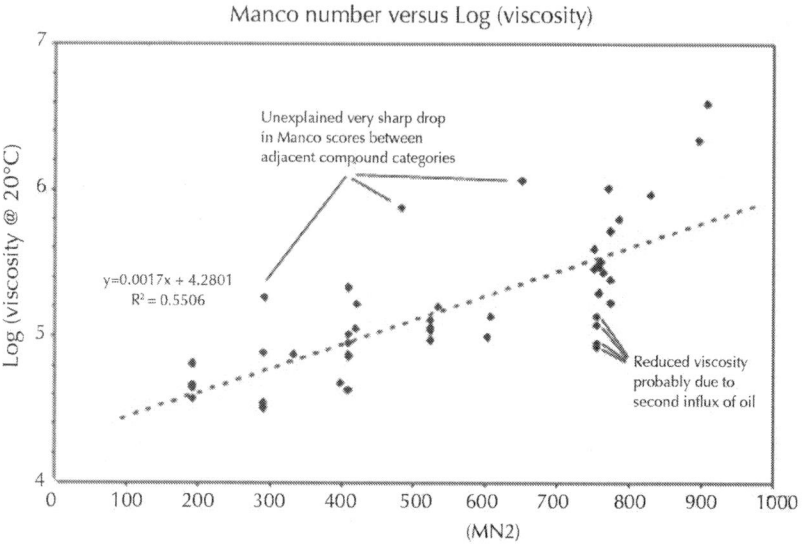

Figure 5: Relationship between measured viscosity and Manco number 2.

MANCO NUMBER INTERPRETATION

The Manco *number* (MN2) is derived from a Manco *vector* of Manco *scores* for a series of progressively more resistant compounds or compound classes. Irregularities in the Manco vector (apparent out of sequence degradation trends) are sometimes observed whereby a normally more labile compound class (e.g. low MW *n*-alkanes or alkyl toluenes) is present in an oil where more refractory compound classes such as methyl naphthalenes or methyl phenanthrenes have been selectively altered or removed by biodegradation. Two or more charging source rocks or mixing of two oil charge volumes could yield an "out of sequence" Manco vector, but it could also potentially be the result of unusual oil chemistry and/or an unusual biological community responsible for a "non-typical" biodegradation process. Anomalous oil chemistry or biological consortia are expected to be uncommon and Occam's razor and the ubiquitous occurrence of oil mixing suggests that out of sequence distributions would normally be interpreted to indicate that a second influx of non-biodegraded oil had been added to a reservoir containing previously biodegraded oil. Subsequent biodegradation of the mixture may ensue as well. Well constrained Manco vectors from a wide variety of geographic and stratigraphic occurrences of heavy oil and bituminous sands may, however, provide sufficient empirical observations to help isolate heretofore unknown controls on the progression of subsurface biodegradation. Clearly the scatter in the relationship between viscosity and Manco number (Fig. 5) suggests that there is not a simple relationship between these parameters.

Processes other than biodegradation can be important in controlling reservoired oil properties. The initial composition of the oil and any secondary charge, water washing under extreme and unusual conditions, and loss of light ends from a heavy oil could result in a significant increase in viscosity, independent of the level of biodegradation. Thermal cracking of heavy oil or bitumen in a reservoir is not anticipated to be a significant process in biodegraded oil reservoirs (Wilhelms et al., 2001), but any process that results in addition of low MW components that would act as a solvent reduces the observed viscosity. Heavy oil occurrence in reservoirs >80 °C is extremely unusual but does occur in complex tectonic settings. The most likely source of light end enriched oils in heavy oil settings is simply later oil charge (Adams et al., 2010). Similarly, natural gas deasphalting caused by the introduction of methane into an undersaturated light oil reservoir could in principle result in a separation of a light, lower viscosity oil from a separated, more viscous, more asphaltene rich oil phase, essentially independent of the extent of biodegradation but, in general, these situations are not commonly observed or expected. Variation in oil charge, including mixing of multiple sourced oils or multiple maturity oil charges, is the principal control on oil viscosity and API gravity for heavy oils, in addition to oil biodegradation.

INTEGRATION OF CONCENTRATION DATA AND MANCO SCORES TO ASSESS PROCESSES AND OIL MIXING

Fig. 6 shows the relationship between the absolute concentration (ppm) of alkyl toluenes (AT), C_2 and C_3 alkyl naphthalenes (C2N, C3N), phenanthrene, the 9-methylphenanthrene/1-methyl phenanthrene ratio and the Manco number (MN2) for a large set of contiguous oils. All fractions show a gradual and progressive removal of components (points A, B, C) with degradation (increasing Manco number). Alkyl toluenes are degraded at lower Manco number, but each fraction is degraded continuously over a wide range of Manco numbers, as indicated by decreasing concentration of each fraction. Recent charge is indicated by a high concentration of a component at relatively high Manco number. Methyl phenanthrenes are relatively stable to biodegradation, as seen by the cross plot of 9MP/1MP where alkyl phenanthrene isomer ratios do not change until MN2 reaches ca. 750, but phenanthrene itself is clearly being removed, from Manco number 400, and disappears over a wide range of biodegradation level. As the Manco number is biased towards the presence of the most degraded oil fraction, fresh oil charges are indicated by an anomalously high concentration of components at a given Manco number. The Manco number approach allows visualization of biodegradation trends and assessment of degradation level and is a useful geochemical tool for selection when key isomer ratio changes occur.

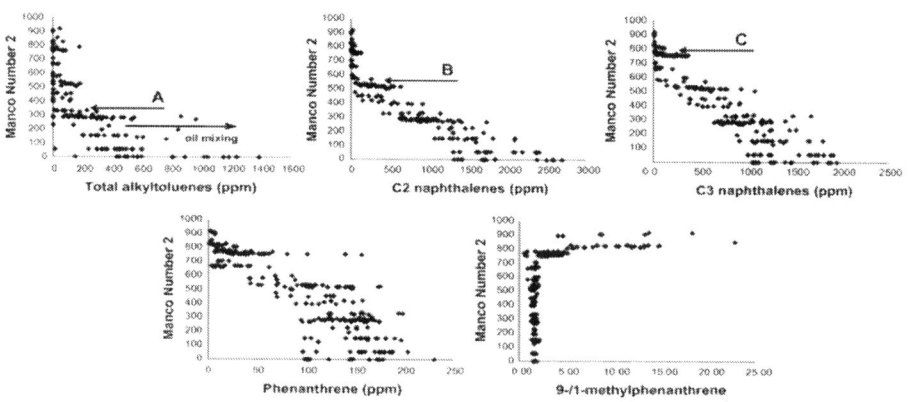

Figure 6: Relationship between absolute concentration (ppm) of alkyl toluenes, C_2 and C_3 alkyl naphthalenes, phenanthrene and 9-methylphenanthrene/1-methylphenanthrene (9-MP/1-MP) vs. Manco number 2 (MN2). All fractions show a gradual and progressive removal of components (points A, B and C). Alkyl toluenes are degraded at lower MNs, but each fraction is degraded over a wide range of MN, as indicated by the concentration of each fraction. Recent charge is indicated by a high concentration of a component at a relatively high MN. Methyl phenanthrenes are relatively resistant, as seen from the cross plot of 9-MP/1-MP where alkyl phenanthrene isomer ratio values do not change until MN2 reaches ca. 750, but phenanthrene itself is clearly removed by MN 400. As MN is biased to the presence of the most degraded oil fraction, a fresh oil charge is indicated by an anomalously high concentration of components at a given MN [after Adams (2008)]. Note: Data for six related samples from one reservoir compartment are not plotted as they showed anomalous concentrations of *some* aromatic hydrocarbons presumed to be contaminants.

CLASSIFYING OILS AT PM LEVEL 0–4

A similar approach can be used for levels of biodegradation lower than those for heavy oil and oil sand deposits. Here the pertinent classes consist of Class 0 [low MW n-alkanes ($<C_{15}$)], Class 1 [high MW n-alkanes ($>C_{15}$)] and Class 2 (isoprenoid alkanes). Two new Manco numbers are then defined for light and moderate levels of biodegradation as follows.

Light and Moderate Biodegradation Manco Numbers

Equations for calculating Manco numbers for light to moderately degraded oil are analogous to those shown in Section 2.1 except that only three compound classes are used rather than eight.

Manco number 1 : MN1 $= \Sigma(\text{Class score}_i \times 5^i)$ for classes i

$$= 0{-}2 \tag{3}$$

Manco number 2 : MN2

$$= [(\text{number of compound classes})$$
$$+ (\log_5(\text{MN1}) \times (\text{scale maximum}$$
$$- 1)]/(\text{number of compound classes}) \tag{4}$$

If MN1 $= 0$, then MN2 is assigned a value of 0. The spreadsheet formula for evaluating MN2 is thus: Manco number 2: MN2 $=$ IF{MN1 > 0 then [3 + (999 * log[MN1, 5])]) else 0}/3.

The complete biodegradation level for oils can then be expressed in several ways:

(i) By reporting both light and heavy biodegradation Manco numbers as a couple, e.g. light Manco number, heavy Manco number = 1000, 250 [PM 5] or light Manco number, heavy Manco number = 350, 0 [PM 2]) or light Manco number, heavy Manco number = 500, 450 [PM 5 with light, late charge mixing]; or (ii) by reporting a single combined ultimate Manco number (UMN) by defining a larger biodegradation vector as described below.

This second approach ranks the level of alteration of 11 different compound classes from key mass chromatograms, as before, by simple visual assessment scored into five levels from 0–4. The 11 compound classes consist of class 0 [low MW n-alkanes ($<C_{15}$)];

class 1 [high MW n-alkanes ($>C_{15}$)]; Class 2 (isoprenoid alkanes); plus the eight classes described above and shown in Table 1, i.e. 1-, 2- and 3-ring aromatic hydrocarbons (alkyl toluenes, alkyl naphthalenes and alkyl phenanthrenes), methyldibenzothiophenes and steranes.

Ultimate Manco Number

Again, the calculations are exactly analogous to those shown above except that the number of compound classes is now 11 rather than 8 or 3.

Ultimate Manco number 1 : UMN1 $= \Sigma(\text{Class score}_i * 5^i)$
$$\text{for classes } i = 0-10 \tag{5}$$

Ultimate Manco number 2 :IF $\{MN1 > 0$ then [
$$11 + (999 * \log[MN1, 5])] \text{ else } 0\}/11 \tag{6}$$

The scale can be extended above PM 8 by including hopanes, 25-norhopanes, aromatic steroid hydrocarbons and diamondoid hydrocarbons and so on. While the principle is clear, increasingly complex behavior in that high MW range has been observed and will be reported in a future study. The behavior of non-hydrocarbons is probably more useful in the post PM 8 range.

CONCLUSIONS

The relative extent of biodegradation of heavy oil and oil sand bitumen may be described in terms of a rigorously defined parameter, the Manco number. The parameter is designed to be useful to detect compositional and thus fluid property differences that are too subtle to be differentiated using earlier biodegradation scales (Peters and Moldowan, 1993, Wenger et al., 2002 and Peters et al., 2005). The latter are too insensitive and lack discrimination within the heavy and severe levels of biodegradation commonly seen in heavy oil provinces. The Manco scale is applicable between PM values of 4–8 and can allow recognition of oils containing a mixture of more biodegraded oil and a lighter, typically later oil charge. This new scale adequately differentiates oils near the transition between reservoirs that are cold producible and those for which thermal stimulation is necessary to recover oil. While there is a general correlation between Manco number and viscosity, the relationship is not particularly strong because of factors other than biodegradation that may influence oil viscosity. An expanded ultimate Manco number is also defined which effectively replaces the PM scale from pristine oil (PM 0) up to PM 8.

ACKNOWLEDGMENTS

The authors thank Gushor Inc. and the Petroleum Reservoir Group at the University of Calgary for permission to publish. We are grateful to two anonymous reviewers who made many excellent suggestions on how to improve the manuscript.

REFERENCES

1.   Adams, J.J., 2008. The Impact of Geological and Microbiological Processes on Oil Composition and Fluid Property Variations in Heavy Oil and Bitumen Reservoirs. PhD Thesis. University of Calgary.

2.   Adams, J.J., Riediger, C., Fowler, M., Larter, S.R., 2006. Thermal controls on biodegradation around the Peace River tar sands: paleo-pasteurization to the west. Journal of Geochemical Exploration 89, 1–4.

3.   Adams, J.J., Jiang, C., Bennett, B., Huang, H., Oldenburg, T., Noke, K.J., Snowdon, L.R., Gates, I.D., Larter, S.R., 2008. Viscosity determination of heavy oil and bitumen: caution and solutions. World Heavy Oil Congress: Unconventional Oil Challenging Conventional Expectations, March 10–12. Shaw Conference Centre, Edmonton, Alberta, Canada (Paper 2008-443).

4.   Adams, J.J., Larter, S.R., Bennett, B., Marcano, N.I., Oldenburg, T.B.P., 2010. Alberta oil sands charge allocation mapping source rock contributions. In: American Association of Petroleum Geologists International Conference & Exhibition – Frontiers of Unconventional Thinking: Saddle Up for the Ride. TELUS Convention Centre, Calgary, Alberta, Canada. <http://www.searchanddiscovery.   com/abstracts/pdf/2010/intl/abstracts/ndx_adams.pdf>.

5.   Bennett, B., Larter, S.R., 2000. Quantitative separation of aliphatic and aromatic hydrocarbons using silver ion-silica solid-phase extraction. Analytical Chemistry 72, 1039–1044.

6.   Bennett, B., Larter, S.R., 2008. Biodegradation scales: applications and limitations. Organic Geochemistry 39, 1222–1228.

7.   Bennett, B., Aitken, C.M., Jones, D.M., Larter, S.R., 2005. Indicators of Anaerobic Hydrocarbon Degradation in Petroleum Reservoirs. The 15th V.M. Goldschmidt – Petroleum Processes from Source to Trap, May 20–25. Moscow, Idaho, United States, p. A500.

8.   Creaney, S., Allan, J., Cole, K.S., Fowler, M.G., Brooks, P.W., Osadetz, K.G., Macqueen, R.W., Snowdon, L.R., Riediger, C.L., 1994. Petroleum generation and migration in the Western Canada Sedimentary Basin. In: Mossop, G.D., Shetsen, I. (Eds.), Geological Atlas of the Western Canada Sedimentary Basin (Chapter 31). <http://www.ags.gov.ab.ca/publications/wcsb_atlas/atlas.html>.

9.   Head, I.M., Jones, D.M., Larter, S.R., 2003. Biological activity in the deep subsurface and the origin of heavy oil. Nature 426, 344–352.

10.   Horstad, I., Larter, S.R., 1997. Petroleum migration, alteration, and remigration within Troll field, Norwegian North Sea. American Association of Petroleum Geologists Bulletin 81, 222–248.

11.   Jiang, C., Bennett, B., Larter, S.R., Adams, J.J., Snowdon, L.R., 2010. Viscosity and API gravity determination of solvent

extracted heavy oil and bitumen. Journal of Canadian Petroleum Technology 49, 20–27.

12.   Koopmans, M.P., Larter, S.R., Zhang, C., Mei, B., Wu, T., Chen, Y., 2002. Biodegradation and mixing of crude oils in Eocene Es3 reservoirs of the Liaohe basin, northeastern China. American Association of Petroleum Geologists Bulletin 86, 1833–1843.

13.   Larter, S.R., Wilhelms, A., Head, I., Koopmans, M., Aplin, A., di Primio, R., Zwach, C., Erdmann, M., Telnaes, N., 2003. The controls on the composition of biodegraded oils in the deep subsurface – Part 1: biodegradation rates in petroleum reservoirs. Organic Geochemistry 34, 601–613.

14.   Larter, S.R., Gates, I., Adams, J., Bennett, B., Huang, H., Koksalan, T., Fustic, M., 2006a. Reservoir fluid characterization of tar sand and heavy oil reservoirs-impact of fluid heterogeneity on production characteristics. American Association of Petroleum Geologists 2006 Annual Convention – Perfecting the Search – Delivering on Promises, April 9–12. George R. Brown Convention Center, Houston, Texas, USA.

15.   Larter, S.R., Huang, H., Adams, J.J., Bennett, B., Jokanola, O., Oldenburg, T.B.P., Jones, M., Head, I.M., Riediger, C.L., Fowler, M.G., 2006b. The controls on the composition of biodegraded oils in the deep subsurface. Part II – geological controls on subsurface biodegradation fluxes and constraints on reservoir-fluid property prediction. American Association of Petroleum Geologists Bulletin 90, 921–938.

16.   Larter, S.R., Adams, J., Gates, I.D., Bennett, B., Huang, H., 2008a. The origin, prediction and impact of oil viscosity heterogeneity on the production characteristics of tar sand and heavy oil reservoirs. Journal of Canadian Petroleum Technology 47, 52–61.

17.   Larter, S.R., Gates, I.D., Adams, J.J., Jiang, C., Snowdon, L.R., Bennett, B., Huang, H., 2008b. Preconditioning an oilfield reservoir. Patent WO/2008/070990.

18.   Larter, S.R., Bennett, B., Snowdon, L.R., Jiang, C., Adams,

J.J., Gates, I.D., Noke, K.J., 2009. Method for determining a value of a property of oil extracted from a sample. Patent WO/2009/023953.

19. Pepper, A.S., 1991. Estimating the petroleum expulsion behaviour of source rocks: a novel quantitative approach. In: England, W.A., Fleet, A.J. (Eds.), Petroleum Migration, vol. 59. Geological Society, Special Publications, London, pp. 9–31.

20. Peters, K.E., Moldowan, J.M., 1993. The Biomarker Guide: Interpreting Molecular Fossils in Petroleum and Ancient Sediments. Prentice Hall, Englewood Cliffs, NJ. Peters, K.E., Walters, C.C., Moldowan, J.M., 2005. The Biomarker Guide 2. Biomarkers and Isotopes in Petroleum Exploration and Earth History. Cambridge University Press, New York.

21. Volkman, J.K., Alexander, R., Kagi, R.I., Woodhouse, G.W., 1983. Demethylated hopanes in crude oils and their applications in petroleum geochemistry. Geochimica et Cosmochimica Acta 47, 785–794.

22. Wardlaw, G.D., Nelson, R.K., Reddy, C.M., Valentine, D.L., 2011. Biodegradation preference for isomers of alkylated naphthalenes and benzothiophenes in marine sediment contaminated with crude oil. Organic Geochemistry 42, 630– 639.

23. Wenger, L.M., Davis, C.L., Isaksen, G.H., 2002. Multiple controls on petroleum biodegradation and impact on oil quality. SPE Reservoir Evaluation and Engineering 5, 375–383.

24. Wilhelms, A., Larter, S.R., Head, I., Farrimond, P., di Primio, R., Zwach, C., 2001. Biodegradation of oil in uplifted basins prevented by deep-burial sterilization. Nature 411, 1034–1037.

Chapter 2

Experimental Study of Micro-particle fouling under Forced Convective Heat Transfer

S. M. Peyghambarzadeh[I], A. Vatani[I], and M. Jamialahmadi[II]

[I]Department of Chemical Engineering, College of Engineering, University of Tehran, Tehran, Iran
[II]Petroleum University of Technology, Ahvaz, Iran

ABSTRACT

Particulate fouling studies of a hydrocarbon based suspension containing 2 μm alumina particles were performed in an annular heat exchanger having a hydraulic diameter of 14.7 mm. During fouling experiments, the classical asymptotical behavior was observed. It is shown that particle concentration, fluid velocity, and wall temperature have strong influences on the fouling curve and the asymptotic fouling resistance. Furthermore, a mathematical model is developed to formulate the asymptotic fouling resistance in terms of the mass transfer coefficient, thermophoresis velocity, and fluid shear rate. The results demonstrate that the prediction of the new model is in good agreement with the experimental observations.

INTRODUCTION

Particulate fouling is defined as the deposition of unwanted material on a heat transfer surface. Products of fouling (i.e., sticking deposits) cause heat transfer resistance and lead to increased capital and maintenance costs and major production and energy losses in many especially energy-intensive industries (Müller-Steinhagen, 2011). The deposition of particulate matter causes problems in many technical applications, e.g., fouling of heat exchangers, contamination of nuclear reactors or blockage of membrane filters, leading to a wide-spread interest in the development of methods to predict and control the rate at which fine particles, suspended in liquids, deposit on walls (Adomeit and Renz, 1996). The subject of heat transfer in particulate liquids became popular during the 1950s (Ozbelge and Koker, 1996) and up to now several theoretical and experimental investigations have been performed. Some of the important published papers, especially experimental studies in the area of particulate fouling, are cited here.

Epstein (1997) stated that particulate fouling is unlike most other categories of fouling. It shows no delay time before deposition is recorded, commonly yields a plot of mf (or Rf) vs t that follows a falling rate, with mf (or Rf) approaching an asymptotic value. He studied and summarized the particle deposition from suspensions flowing parallel to nonporous smooth and rough surfaces in terms of particle transport to, attachment at, and reentrainment from the surface. Particle deposition is commonly considered to be a two-step process: a transport step, in which particles are transferred to the wall, and a subsequent adhesion step, which is dominated by the interaction forces between particles and wall (Adomeit and

Renz, 1996). Most of the investigations into transport mechanisms have concentrated on the situation in gas flow, where transport is frequently dominated by inertial effects (Papavergos and Hedley, 1984). Investigations concerning particle transport in liquids predominantly accounted for Brownian diffusion (Bowen et al., 1976), in which the transport rate can be calculated analytically for laminar flow. In turbulent flow, empirically based correlations, e.g., that suggested by Metzner and Friend (1958), are used or theories for aerosol deposition are adapted. Since in liquids the particle relaxation time is considerably smaller than in gases, the theories that account for diffusive effects are applicable, such as those of Beal (1970) and Davies (1983).

Earlier experimental investigations on particle deposition have focused on the influence of chemical materials on the deposition rates in isothermal laminar (Bowen and Epstein, 1979) and turbulent flow (Williamson et al., 1988). Other influences on deposition rates, such as temperature, heat transfer rate and flow rate have been investigated in less detail. Due to the practical importance of these influences, there is a strong need for data obtained under systematically varied adhesion conditions, covering a wide range of chemical, thermal and hydrodynamic parameters.

LITERATURE REVIEW

Williamson et al. (1988) studied the deposition of haematite ($-Fe_2O_3$) particles of 0.2 µm diameter in suspension in water and found that the deposition is crucially dependent on the suspension pH. Melo and Pinheiro (1988) carried out particulate fouling tests using kaolin-water suspensions flowing through an annular heat exchanger. They used these data in a comparative study of several transport models. It was found that, in the lower range of fluid velocities (less than 0.5 m/s), the deposition seemed to be controlled by mass transfer. Kim and Webb (1991) developed a fouling model with an experimentally determined sticking probability and deposit bond strength factor. Their model could predict the fouling behavior of repeated rib tubes. The mass transfer rate was assumed to control the particle transport process and the wall shear stress assumed to control the removal process. The Reynolds number was between 14000- 26000 and the foulant particles contained ferric oxide and aluminum oxide. Furthermore, an analysis was performed that accounts for the forces acting on the particles at the wall. The deposition rate of Lake Ontario slit onto type 304 stainless steel was given by Turner and Lister (1991). They found that only particles less than 5 µm were found in the deposit, even though particles up to 25 µm were in the flowing slit suspension.

The deposition of fine silica and polystyrene spheres was measured for conditions of laminar and turbulent flow in a rectangular channel using image analysis. Contrary to the results for laminar flow, the initial deposition rates in the turbulent flow decreased with increasing Reynolds number, indicating that deposition was no longer mass transfer controlled (Vasak et al., 1995). Karabelas et al. (1997) reported particulate fouling data for plate heat exchangers with particles of mean size 5 µm. The flow passage geometry and the fluid velocity had a strong effect on the fouling resistance. These results showed that fouling was adhesion controlled and that the maximum measured resistance was almost an order of magnitude smaller than the TEMA recommendations. They reported that tangential hydrodynamic forces are responsible for particle detachment from the heat transfer surface.

Grandgeorge et al. (1998) performed an experimental study on the liquid-phase particulate fouling of stainless steel corrugated plate heat exchangers. Deionized water containing TiO_2 particles was used as the foulant fluid. During fouling experiments, asymptotical behavior was observed. A systematic study of the influence of the fluid velocity on the initial deposition rate as well as on the asymptotic thermal resistance of the deposit was performed. The influence of the suspension pH on the fouling process was also provided. They also proposed a falling-rate global model based on the idea that the adhesion coefficient decreases when the particles accumulate on the wall. Turner and Klimas (2000) measured the deposition rate of colloidal magnetite particles at alkaline pH under both single phase forced convection and flow boiling conditions. The deposition of magnetite particles from suspension in water at 90 °C was studied under various conditions of flow, chemistry, and boiling heat transfer. The experiments indicated that, under non-boiling conditions, mechanisms based on diffusion and thermophoresis control deposition, while removal is negligible (Basset et al., 2000). Coutinho et al. (2001) described a thermodynamic model for predicting the deposition of paraffinic waxes during production in hydrocarbon fluids at low and high pressures. Yiantsios and Karabelas (2003) obtained a set of experimental data for micrometer-sized particle deposition under well- controlled hydrodynamic conditions covering a range of physicochemical conditions, particle sizes and substrate materials. Their experimental results revealed the important effects of gravity, lift forces, and physicochemical interactions. Even for particles in the micrometer-size range, the sticking probability was limited by hydrodynamic conditions that are similar to or less severe than those encountered in industrial heat exchangers.

Buchelli et al. (2005) analyzed particulate fouling in a continuous LDPE polymerization reactor and found that the foulant thickness grew linearly with time. Based on the heat and mass transfer analogy and analysis of the plant data they suggested that only a small fraction of the polymer that is precipitated near the reactor wall gets attached to the wall to produce fouling. Li (2007) provided data for accelerated particulate fouling in helically ribbed copper tubes at different concentrations, velocities, and geometries. Aluminum oxide particles with 3 µm average particle diameters were used as foulant. The semi-theoretical analysis of dividing a fouling factor ratio into a fouling process index and an efficiency index significantly simplified the fouling analysis.

Recently, Jamialahmadi et al. (2009) studied the mechanisms of deposition of flocculated asphaltene particles from oil experimentally and theoretically under forced convective conditions using an accurate thermal approach. It was observed that, during the first few weeks the deposition mechanism was dominant and the erosion of the deposit was almost negligible. The rate of asphaltene deposition increased with increasing flocculated asphaltene concentration and temperature, while it decreased with increasing oil velocity.

In this paper, micron-sized α-alumina particles are introduced into a hydrocarbon base fluid (n-heptane). There is almost no information about the deposition from hydrocarbon-based suspensions in the literature. The deposition during forced convection heat transfer was measured using an accurate thermal approach. Furthermore, by analyzing the effect of operating conditions, a new theoretical model was developed to predict the asymptotic fouling resistance. Results of this study could be applied in different industrial cases such as crude oil refineries and various types of water-cooled heat exchangers in chemical, food processing and power plants.

EXPERIMENTAL

Experimental Apparatus and Materials

Figure 1 shows the test apparatus used for the present investigation. The solution flows in a closed loop consisting of a temperature controlled storage tank, a centrifugal pump and an annular test section. The flow velocity of the solutions was measured with a calibrated flow meter (Technical Group LZM-15Z Type) with the accuracy of ±0.1 l/min. The fluid temperature was measured by two RTDs (Pt-100Ω) located in mixing chambers before and after the test section. The complete system was made from stainless steel. Thermocouple voltages and the current and voltage drop from the test heater were measured and processed with a data acquisition system in conjunction with a computer. The solution temperature was controlled by cooling water coils inside the tanks and by electrical band heaters in conjunction with temperature controllers.

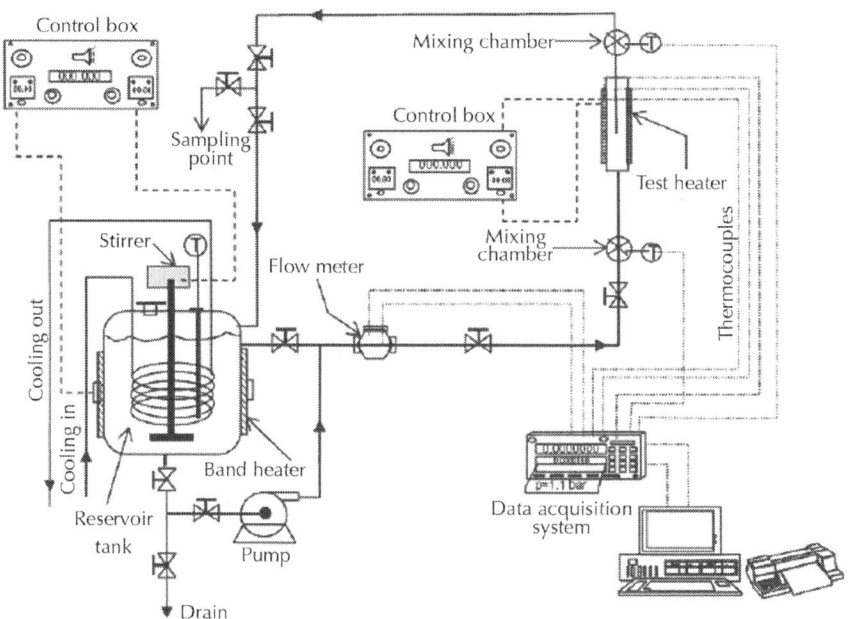

Figure 1: Simplified diagram of the experimental apparatus.

When the fluid is heated in the annular space, alumina micro-particles are deposited on the test heater. This heater was manufactured by the Ashland Chemical Company (One Drew Plaza Boonton, NJ 07005, US) according to an HTRI design. Inserting the test heater in the flow line creates an annular section. In the test section, the local wall temperatures were measured with four stainless steel sheathed miniature thermocouples (E-type), which were installed as close as possible to the heat transfer surfaces. The simplified scheme of the test heater, with the thermocouples' location in it, is demonstrated in Figure 2. This type of heater has been extensively used for fouling research by the other investigators (Najibi et al., 1997; Helalizadeh et al., 2000). The temperature drop between the thermocouple location and the heat transfer surface can be calculated from:

$$T_w = T_{th} - q\frac{s}{\lambda_w}$$

(1)

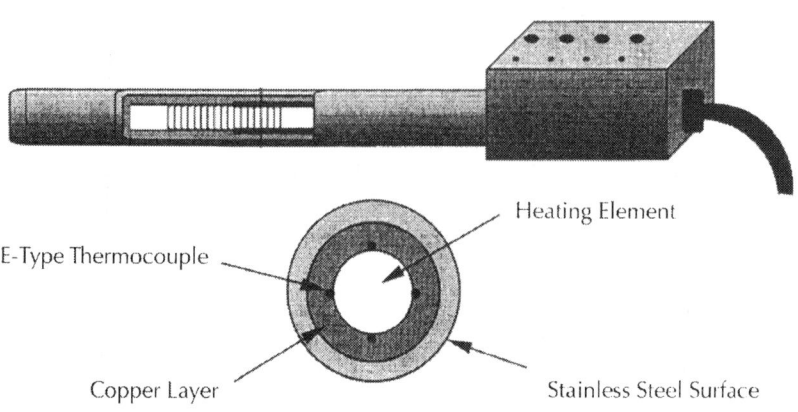

E-Type Thermocouple

Heating Element

Copper Layer

Stainless Steel Surface

Figure 2: Schematic of the test heater.

The ratio between the distance of the thermocouples from the surface and the thermal conductivity of the tube material (s/λ_w) was determined for each thermocouple by calibration measurements using a Wilson plot technique (Fernández-Seara et al., 2007). The average temperature difference for each test section was the arithmetic average for the four thermocouple locations around the rod circumference. The average of the voltage readings was used to determine the difference between the wall and bulk temperature for each thermocouple. All thermocouples used were thoroughly calibrated by using a constant temperature water bath, and their accuracy has been estimated to be ±0.2 °C. The local heat transfer coefficient α is then calculated from

$$\alpha = \frac{q}{T_w - T_b}$$

(2)

The important dimensions of the test section are shown in Table 1.

Table 1: Dimensions of the test section

Dimensions	Value (mm)
Heater diameter	10.7
Annulus outside diameter	25.4
Heated length	99.1
Hydraulic diameter	14.7
Heated length to thermocouple location	82.6
Entrance length to heater beginning	216.0

Prior to commencing a test run, the test heater, reservoir tanks and pipes were washed and cleaned with water and cleaning agent to remove any scale from previous experiments. After the system was cleaned, the test solution was introduced into the reservoir tanks. Following this, the tank heater was switched on and the temperature of the system allowed to rise. Once the fluid reached the desired temperature, the pump was started and the system left to stabilize to the desired bulk temperature and velocity. Then power was supplied to the test heater and kept at a pre-determined value. The data acquisition system was switched-on to record temperatures, pressure and heat flux.

The base liquid was a pure hydrocarbon (n-heptane). It was selected since there are a large number of investigations in the literature on the deposition of fine particles from water-based suspensions. In addition, its boiling point and visual characteristics are close to those of water. Some of the important physical properties of n-heptane at 20 °C are shown in Table 2. Al_2O_3 microparticles were added to the base fluid as the foulant at different concentrations including 0-200 mg/l. In this study, alumina particles were selected since it is chemically inert in relation to the base fluid and there will be no interference of chemical reaction fouling in the experiments. The particle specifications are demonstrated in Table 3.

Table 2: Physical properties of n–heptane at 288.15 K (Campbell, 1992)

ρ	μ	k	C_p	M	P_c	T_c	T_{NBP}
(kg/ m³)	(kg/m.s)	(W/m.K)	(kJ/ kg.K)	(kg/ kmol)	(kPa)	(K)	(K)
688	$4.1*10^{-4}$	0.14	2.209	100.2	2736	540.2	371.6

Table 3: Physical properties of α–alumina at 20 °C (Barin and Knacke, 1973; Kuzmann, 1976; Karabelas et al., 1997)

M	k	C_p	ρ	$d_{p, ave}$
(kg/kmol)	(W/m.K)	(kJ/kg.K)	(kg/m³)	(μm)
102	3.31	656.3	3990	2

In all of the experiments, 250 ml of the circulating fluid was periodically withdrawn from the drain line of the storage tank (see Fig. 1). This sample was passed through a filter paper (1 μm pore diameter) and the filter paper then dried in an oven and weighed. The weight difference demonstrated that the reduction in the particle concentration during the experiments was negligible in comparison with the initial particle concentration. Therefore, the particle concentration was constant over the whole experiment. It should also be mentioned that no dispersant or stabilizer was added to the suspension. This is due to the fact that the addition of any agent could change the fluid properties and consequently, might influence its deposition behavior. Furthermore, creating a turbulent flow condition in the experiments guarantees the stabilization of the nanoparticles in nheptane. The range of operating parameters is shown in Table 4.

Table 4: Range of operating conditions used in this work

Condition	Range
q (W/m²)	Zero – 70000

T_w (K)	300 – 350
T_b (K)	300 – 330
Re	6000 – 18000
u (m/s)	0.2 – 0.7
C_b (kg/m³)	Zero – 0.2

Error Analysis

Uncertainty analysis was carried out by calculating the error of the measurements. The uncertainty range of the Reynolds number comes from the errors in the measurement of the volumetric flow rate and hydraulic diameter of the tubes as follows:

$$Re = \frac{\rho u d}{\mu} = \frac{4\rho \dot{V}}{\mu \pi d_h} \tag{3}$$

$$\left[\frac{\delta(Re)}{Re}\right]^2 = \left[1 \times \frac{\delta \dot{V}}{\dot{V}}\right]^2 + \left[(-1) \times \frac{\delta d_h}{d_h}\right]^2 \tag{4}$$

The results demonstrate that the main source of uncertainty in the calculation of Reynolds number was due to the measurement of the volumetric flow rate.

The uncertainty of the heat transfer coefficient (or equally, fouling resistance) refers to the errors in the measurements of the volumetric flow rate, hydraulic diameter, and all the temperatures.

$$\alpha = \frac{mC_p(T_{in} - T_{out})}{A(T_w - T_b)} = \frac{\rho \dot{V} C_p(T_{in} - T_{out})}{A(T_w - T_b)} \tag{5}$$

$$\left[\frac{\delta \alpha}{\alpha}\right]^2 = \left[1 \times \frac{\delta \dot{V}}{\dot{V}}\right]^2 + \left[(-1) \times \frac{\delta A}{A}\right]^2$$

$$+ \left[1 \times \frac{\delta(T_{in} - T_{out})}{(T_{in} - T_{out})}\right]^2 + \left[(-1) \times \frac{\delta(T_w - T_b)}{(T_w - T_b)}\right]^2 \tag{6}$$

The results demonstrate that the most important source of the uncertainty in the calculation of the heat transfer coefficient is due to the measurement of temperature differences, especially the last term in Eq. (6).

According to the uncertainty analysis described by Moffat (1988), the measurement errors of the main parameters are summarized in Table 5. Furthermore, to check the reproducibility of the experiments, some runs were repeated later, which proved to be excellent.

Table 5: The uncertainties of the measured parameters

Parameter	Value	Unit	Uncertainty
A	3.3×10^{-4}	m^2	0.94%
Re	$6 \times 10^3 - 1.8 \times 10^4$	–	5.2%
α	$2 \times 10^2 - 1.6 \times 10^3$	W/m^2K	10.4%
R_f	Zero $- 4.5 \times 10^{-3}$	m^2K/W	10.4%

RESULTS AND DISCUSSION

It is customary to present fouling data in terms of fouling resistance (R_f), which can be calculated on the basis of the overall heat transfer coefficient at t = 0 and at any desired time:

$$R_f = \frac{1}{\alpha(t)} - \frac{1}{\alpha(t = 0)}$$

$$(7)$$

For the range of the operating variables investigated, an asymptotic increase of fouling resistance with time was observed, which is in accordance with the previous particulate fouling experiments (Kim and Webb, 1991; Grandgeorge et al., 1998; Sheikholeslami, 2000; Bansal et al., 2001; Walker and Sheikholeslami, 2003; Müller-Steinhagen, 2011). The values of fouling resistance were measured under different operating conditions, including particle concentration, fluid velocity, and wall temperature. All of the experimental data presented in this study are summarized in Table 6 for better demonstration of the effect of the operating variables.

Table 6: All of the experimental runs presented in this study and the corresponding results

#	T_b (°C)	T_w (°C)	q (kW/m²)	C_b (mg/l)	u (cm/s)	R_f^* (m²K/kW)
1	28.0	48.2	15	15	33	0.39
2	28.5	48.1	15	20	33	0.65
3	28.8	49.0	15	30	33	0.95
4	29.1	48.9	15	40	33	1.7
5	28.8	48.3	15	55	33	1.8
6	28.5	48.6	15	70	33	2.5
7	28.9	49.4	15	70	20	3.68
8	28.9	49.2	15	70	45	2.21
9	28.1	49.4	15	70	52	1.80
10	27.9	38.6	15	70	65	0.65
11	29.1	31.0	1.5	70	33	0.50
12	28.8	32.8	3	70	33	0.81
13	29.5	37.5	6	70	33	1.80
14	29.0	49.1	25	70	65	1.80
15	28.6	78.1	57	70	33	3.59
16	29.0	49.1	15	80	33	2.61
17	28.7	48.7	15	100	33	2.70
18	28.5	49.0	15	110	33	3.29
19	27.8	47.7	15	170	33	4.30
20	28.4	43.1	15	170	65	0.85

Furthermore, some of the experimental results of this investigation are shown in Figures 3-8. As can be seen, a sharp increase of fouling resistance occurred in the initial period of time. After this time, the gradual increase of fouling resistance indicates the simultaneous effect of deposit removal.

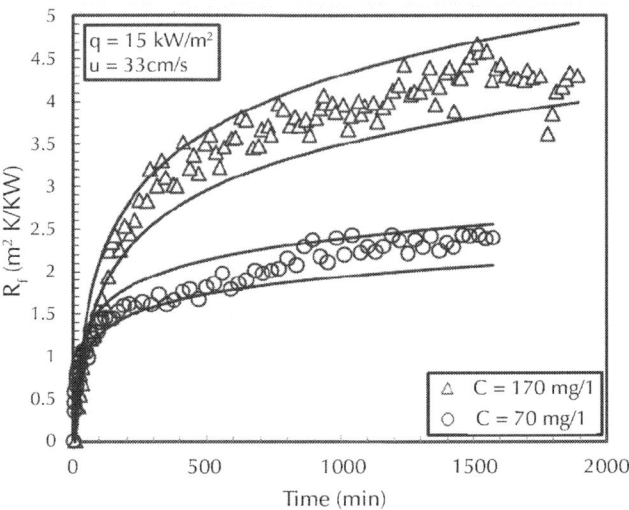

Figure 3: The effect of particle concentration on the fouling curve.

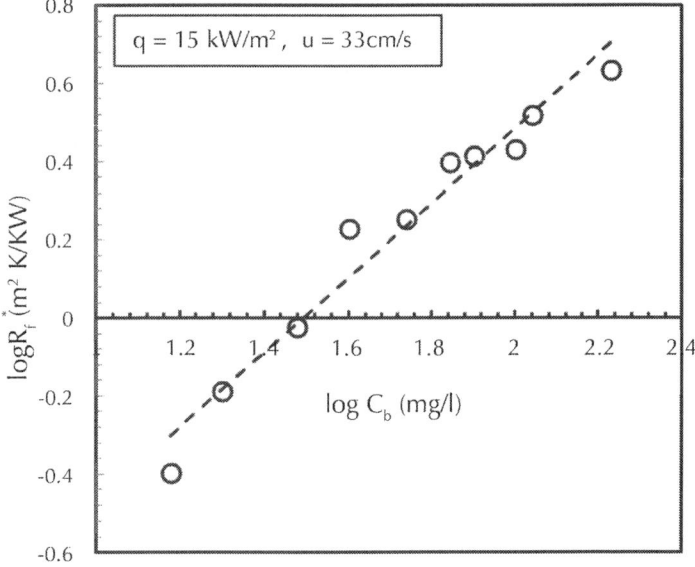

Figure 4: The effect of particle concentration on the asymptotic fouling resistance.

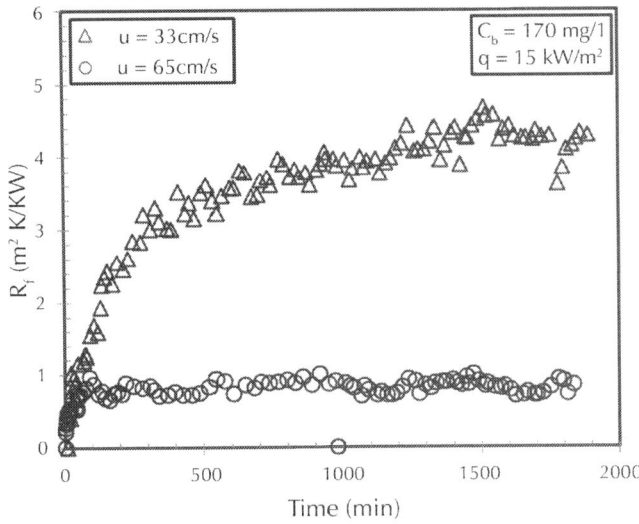

Figure 5: The effect of fluid velocity on the fouling curve.

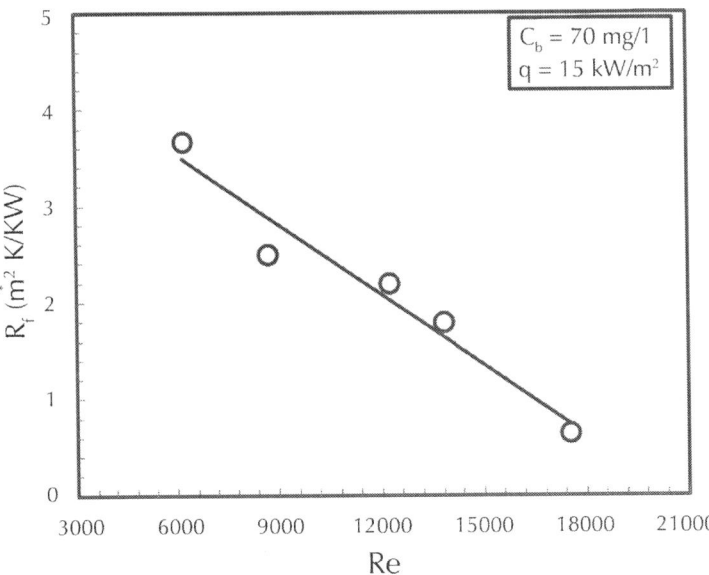

Figure 6: The effect of fluid velocity on the asymptotic fouling resistance.

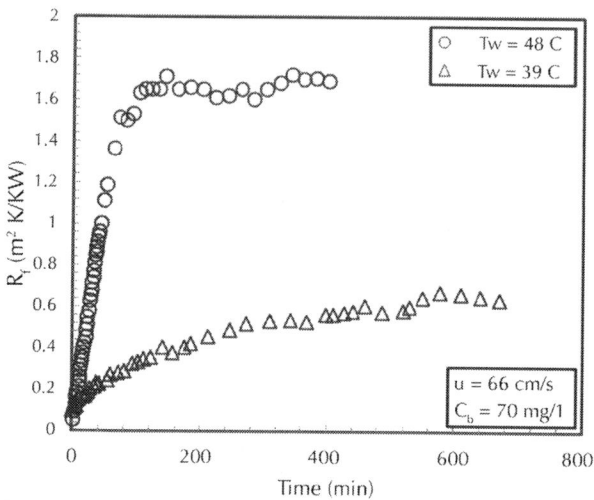

Figure 7: The effect of wall temperature on the fouling curve.

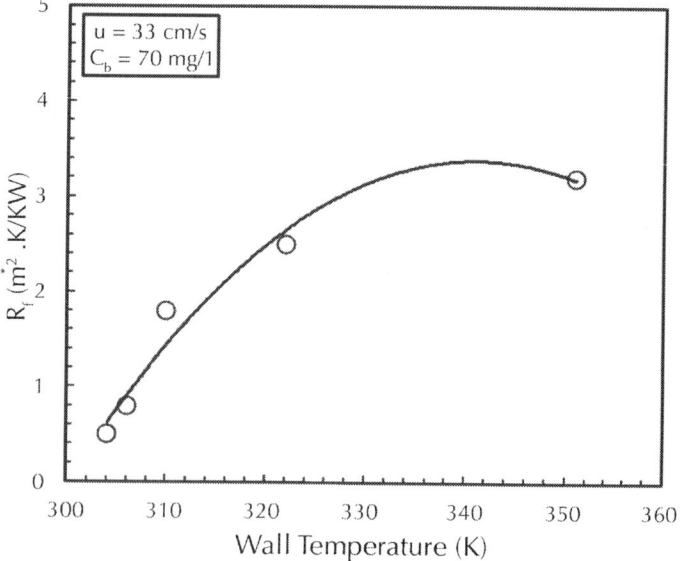

Figure 8: The effect of wall temperature on the asymptotic fouling resistance.

The effect of particle concentration on the fouling resistance is shown in Figure 3. The results show that, as the foulant concentration increases, a higher asymptotic value of fouling resistance is obtained. This is an expected result since, at higher concentration of the particles, the agglomeration and settling of particles on the heat transfer surface increases.

The error limits (±10.4%) for the measured fouling resistances are also shown in Fig. 3 as continuous curves.

Figure 4 shows the effect of particle concentration on the asymptotic fouling resistance. As can be seen, a straight line can be fitted to the experimental data, proving that particle concentration has a linear relationship with the asymptotic fouling resistance. Thus, as shown in Fig. 4, the linear fit of the data gives log (R_f^*) = k + n log (C_b) with the near unit slope of n=0.96. This value is in agreement with that reported by other investigations (see, e.g., Jamialahmadi et al., 2000).

Figure 5 demonstrates the effect of flow velocity on the deposition of Al2O3 particles. Under similar operating conditions of particle concentration and heat flux, greater fouling resistance is obtained at lower velocity. As the flow velocity increases, the shear stress exerted by the fluid on the deposit layer increases and, as a result of this force, deposition is hindered or removed.

The asymptotic values of the fouling resistances are plotted as a function of fluid Reynolds number in Figure 6. The asymptotic fouling resistance decreases significantly as the fluid velocity is increased. As will be shown later, the removal rate, $_r$ increases with increasing shear forces at the wall of the fouled tube (Oliveira et al., 1993; Epstein, 1997). Some studies of particle deposition from flowing water suspension have actually shown that the attachment efficiency of particles to the surface also tended to decrease with the wall shear stress (Adomeit and Renz, 2000; Yiantsios and Karabelas, 2003). Thus, the effect of increasing fluid velocity may involve a dual role of enhancing removal forces as well as inhibiting the attachment of particles on the heat transfer surface.

The effect of wall temperature on the fouling curve is shown in Figure 7. It is demonstrated that higher wall temperature causes the rapid deposition of particles on the heat transfer surface and, consequently, greater asymptotic fouling resistance.

The main effect of the wall temperature on the asymptotic fouling resistance is through the diffusivity of particles and the sticking probability in the deposition process. In general, the mass transfer coefficient is a linear function of temperature, while the sticking probability generally increases exponentially with the wall temperature according to an Arrhenius relationship. The non-linear trend shown in Fig. 8 may indicate that both mechanisms are important with respect to the asymptotic fouling resistance.

CORRELATING THE EXPERIMENTAL DATA

Particulate fouling results from the combined effect of the deposition and removal processes. The deposition process can be divided into two consecutive processes: the transport of particles to the wall and the adhesion of particles at the wall. The net deposition rate is generally expressed as (Kern and Seaton, 1959):

$$\frac{dm_f}{dt} = \Phi_d - \Phi_r$$

$$(8)$$

Where Φ_d is the deposition rate and Φ_r is the removal rate. Each term is proved to be (Watkinson and Epstein, 1970):

$$\Phi_d = S_p \cdot K_m \cdot C_b$$

$$(9)$$

And

$$\Phi_r = K_2 \, m_f \, \tau_w$$

$$(10)$$

Where S_p is the sticking probability, which is defined as a probability that a transported particle will stick to the wall (Watkinson and Epstein, 1970). K_m is the conventional mass transfer coefficient and

τ_w the wall shear stress. Insertion of Eqs. (9) And (10) into Eq. (8) yields the following asymptotic fouling resistance equation, which is known as the Kern and Seaton (1959) correlation:

$$R_f = R_f^* \left(1 - e^{-K_2 \tau_w t}\right)$$

(11)

Where

$$R_f^* = \frac{\Phi_d}{\rho_d . \lambda_d . K_2 . \tau_w}$$

(12)

Where λ_d and ρ_d are the deposit thermal conductivity and deposit density. Eq. (12) shows that the asymptotic fouling resistance can be predicted if the mass transfer coefficient K_m, the wall shear stress τ_w, and the sticking probability S_p are known (assuming that we know the deposit density ρ_d and the deposit thermal conductivityλ_d). In this study, the mass transfer coefficient and the wall shear stress were calculated according to the published relations for smooth tubes. The mass transfer coefficient was evaluated using the friction velocity, u*as follows (Cleaver and Yates, 1975):

$$K_m = 0.084 \frac{u^*}{Sc^{0.67}}$$

(13)

Where u* is friction velocity and can be estimated from:

$$u^* = \sqrt{\frac{\tau_w}{\rho_1}} = u\sqrt{\frac{f}{8}}$$

(14)

In the above equation, the Fanning friction factor, f, can be calculated from the classical Blasius equation for smooth tube flow:

$$f = \frac{0.3164}{Re^{0.25}}$$

(15)

Sc is the liquid phase Schmidt number:

$$Sc = \frac{\mu_1}{\rho_1 D_p}$$

(16)

For submicron particles, the Brownian diffusion coefficient, D_p, can be determined from the Stokes Einstein equation (Einstein, 1956):

$$D_p = \frac{\kappa_B T}{3\pi\mu d_p}$$

(17)

The sticking probability, Sp is defined according to (Watkinson and Epstein, 1970):

$$Sp = \frac{K_1 . e^{-\frac{E}{RT_w}}}{u^2}$$

(18)

As described in the literature, a thermal force moves fine particles down a temperature gradient. Hence, cold walls attract and hot walls repel colloidal particles (Epstein, 1997). The thermophoresis velocity, V_T can be estimated according to Whitmore and Meisen (1977) as follows:

$$V_T = \frac{0.26\mu_1}{2\lambda_1 + \lambda_p} \frac{q}{\rho_1 . T}$$

(19)

The thermophoresis effect is especially important at high heat fluxes and, as inferred from the literature, the constant of 0.26 in Eq. (19) should be assigned a value almost an order of magnitude smaller than 0.26 to match the experimental data (Epstein, 1997). In this investigation, due to the low heat fluxes (less than 70 kW/m²) employed in the experiments, the thermophoresis velocity was on the order of 10^{-7} m/s, almost an order of magnitude smaller than the mass transfer velocity, which was on the order of 10^{-6} m/s. Therefore, there was no need to replace the coefficient of 0.26 with the smaller one since the thermophoresis velocity could be ignored in comparison with the mass transfer velocity. The deposition rate of Eq. (13) can be corrected as follows:

$$\Phi_d = \left[K_m - \frac{V_T}{2} \right].S.C_b$$

(20)

Inserting Eqs. (18) And (20) into Eq. (12):

$$R_f^* = \frac{\left[\dfrac{K_1.e^{-\frac{E}{RT_w}}}{u^2} \right].C_b.\left[K_m - \dfrac{V_T}{2} \right]}{K_2.\rho_d.\lambda_d.\tau_w}$$

(21)

Substituting the definitions of Km, V_T, and $_w$ from Eqs. (13) â (19) and considering all the constants in Eq. (21) as K3, the final relation for calculation of R_f^*

$$R_f^* =$$

$$\frac{K_3.e^{-\frac{E}{RT_w}}.C_b.\left[0.084 \dfrac{u.\sqrt{\frac{f}{8}}}{Sc^{0.67}} - \dfrac{0.26\mu_1}{2\lambda_1 + \lambda_p} \dfrac{q}{2\rho_1 T} \right]}{\dfrac{f}{8}.\rho_1.u^2}$$

(22)

In this equation, the values of K_3 and E, estimated from non-linear regression of the experimental data, are 6.5×10^{14} m³.K/J and 63200 kJ/kmol, respectively. The activation energy obtained is in good agreement with previous particulate fouling studies: for example, Watkinson (1969) reported the activation energy in the deposition of a sand-water suspension to be 60000 kJ/kmol. Turner and Lister (1991), in a study of deposition of silt onto stainless steel, found an activation energy of 25300 kJ/kmol. Turner and Klimas (2000) found the value of 81000 kJ/kmol for the deposition of magnetite particles from water.

Figure 9 shows the model prediction for the calculation of asymptotic fouling resistance. The results show good agreement between the model and all of the experimental data. The new model can predict the asymptotic fouling resistance with an absolute average error of about 19%.

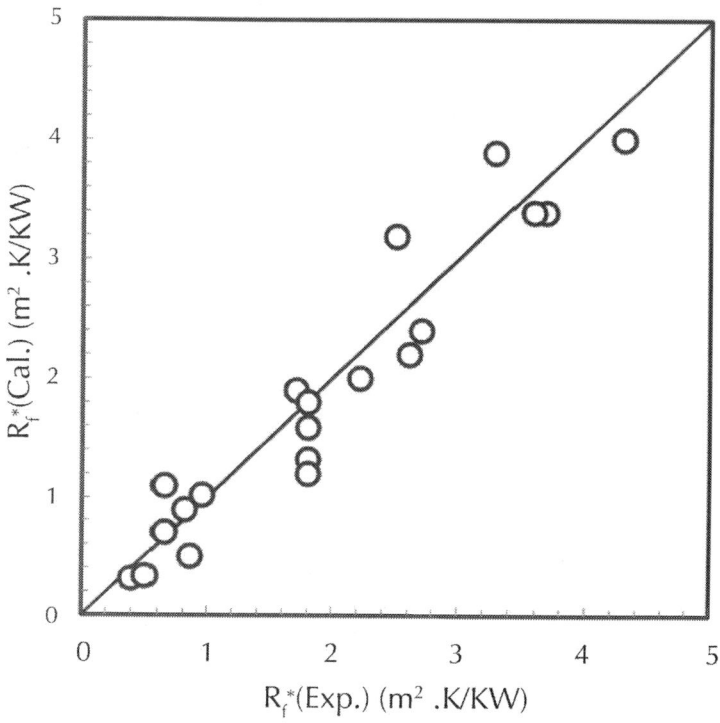

Figure 9: The comparison of the experimental asymptotic fouling resistance with the prediction of our model, Eq. (22).

CONCLUSIONS

This study shows the effects of three key operating parameters, including wall temperature, fluid velocity, and particle concentration, on Al2O3 microparticle fouling. The results show that asymptotic fouling resistance increases with increasing particle concentration and wall temperature and decreasing fluid velocity. Furthermore, a theoretical model is developed for asymptotic fouling resistance that can predict the experimental data with an absolute average error of about 19%. The modeling demonstrates that the effect of thermophoresis is negligible due to the low heat fluxes employed in this investigation.

REFERENCES

1. Adomeit, P., Renz, U., Correlations for the particle deposition rate accounting for lift forces and hydrodynamic mobility reduction. Can. J. Chem. Eng., 78, p. 32(2000).

2. Adomeit, P., Renz, U., Deposition of fine particles from a turbulent liquid flow: experiments and numerical predictions. Chem. Eng. Sci., 51, (13), p. 3491 (1996).

3. Bansal, B., Müller-Steinhagen, H., Chen, X. D., Comparison of crystallization fouling in plate and double-pipe heat exchangers. Heat Transfer Engineering, 22, p. 13 (2001).

4. Barin, I., Knacke, O., Thermochemical Properties of Inorganic Substances. Springer-Verlag, Berlin (1973).

5. Basset, M., McInerney, J., Arbeau, N., Lister, D. H., The fouling of alloy-800 heat exchange surfaces by magnetite particles. Can. J. Chem. Eng., 78, p. 40 (2000).

6. Beal, S. K., Deposition of particles in turbulent flow on channel or pipe walls. Nucl. Sci. Eng., 40, p. 1 (1970).

7. Bowen, B. D., Epstein, N., Fine particle deposition in smooth parallel-plate channels. J. Colloid Interface Sci., 12, p. 81 (1979).

8. Bowen, B. D., Levine, S., Epstein, N., Fine particle deposition in laminar flow through parallel-plate and cylindrical channels. J. Colloid Interface Sci., 54, p. 375 (1976).

9. Buchelli, A., Call, M. L., Brown, A. L., Modeling fouling effects in LDPE tubular polymerization reactors, 1. Fouling thickness determination. Ind. Eng. Chem. Res., 44, p. 1474 (2005).

10. Campbell, J. M., Gas Conditioning and Processing. Volume 1, 7th Edition, Campbell Petroleum Series (1992).

11. Cleaver, J. W., Yates, B., A sublayer model for the deposition of particles from a turbulent flow. Chem. Eng. Sci., 30, p. 983 (1975).

12. Coutinho, J. A. P., Pauly, J., Daridon, J. L., A thermodynamic model to predict wax formation in petroleum fluids. Braz. J. Chem. Eng., 18, (4), p. 411 (2001).

13. Davies, J. T., A new theory of aerosol deposition from turbulent fluids. J. Chem. Eng., 38, p. 135 (1983).

14. Einstein, A., the Theory of Brownian movement. Dover, New York (1956).

15. Epstein, N., Elements of particle deposition onto nonporous solid surfaces parallel to suspension flows. Experimental Thermal and Fluid Science, 14, p. 323 (1997).

16. Fernández-Seara, J., Uhía, F. J., Sieres, J., Laboratory practices with the Wilson plot method. Experimental Heat Transfer, 20, p. 123 (2007).

17. Grandgeorge, S., Jallut, C., Thonon, B., Particulate fouling of corrugated plate heat exchangers. Global kinetic and equilibrium studies. Chem. Eng. Sci., 53, (17), p. 3051 (1998).

18. Helalizadeh, A., Müller-Steinhagen, H., Jamialahmadi, M., Mixed salt crystallisation fouling. Chemical Engineering and Processing, 39, p. 29 (2000).

19. Jamialahmadi, M., Soltani, B., Müller-Steinhagen, H., Rashtchian, D., Measurement and prediction of the rate of deposition of flocculated asphaltene particles from oil. Int. J. Heat Mass Transfer, 52, p. 4624 (2009).

20. Karabelas, A. J., Yiantsios, S. G., Thonont, B., Grillott, J. M., Liquid-side fouling of heat exchangers, an integrated R&D approach for conventional and novel designs. Applied Thermal Engineering, 17, (8-10), p. 127 (1997).

21. Kern, D. Q., Seaton, R. E., A theoretical analysis of thermal surface fouling. Brit. Chem. Eng., 4, p. 258 (1959).

22. Kim, N. H., Webb, R. L., Particulate fouling of water in tubes having a two-dimensional roughness geometry. Int. J. Heat Mass Transfer, 34, (11), p. 2721 (1991).

23. Kuzmann, R., Handbook of Thermodynamic Tables and Charts. Hemisphere Publication Corporation, Washington D.C. (1976).

24. Li, W., Modeling liquid-side particulate fouling in internal helical-rib tubes. Chem. Eng. Sci., 62, p. 4204 (2007).

25. Melo, L. F., Pinheiro, J. D., Particle transport in fouling caused by kaolin-water suspensions on cupper tubes. Can. J. Chem. Eng., 66, p. 36 (1988).

26. Metzner, A. B., Friend, W. L., Theoretical analogies between heat, mass and momentum transfer and modifications for fluids of high Prandtl or Schmidt numbers. Can. J. Chem. Eng., 36, p. 235 (1958).

27. Moffat, R. J., Describing the uncertainties in experimental results. Exp. Therm. Fluid Sci., 1, p. 3 (1988).

28. Müller-Steinhagen, H., Heat transfer fouling: 50 years after the Kern and Seaton model. Heat Transfer Engineering, 32, (1), p. 1 (2011).

29. Najibi, S. H., Müller-Steinhagen, H., Jamialahmadi, M., Calcium sulphate scale formation during subcooled flow boiling. Chem. Eng. Sci., 52, (8), p. 1265 (1997).

30. Oliveira, R., Melo, L., Pinheiro, M., Vieira, M. J., Surface interactions and deposit growth in fouling of heat exchangers. Corrosion Rev., 11, p. 55 (1993).

31. Ozbelge, T. A., Koker, S. H., Heat transfer enhancement in water-feldspar upflows through vertical annuli. Int. J. Heat Mass Transfer, 39, (1), p. 135 (1996).

32. Papavergos, P. G., Hedley, A. B., Particle deposition behaviour from turbulent flows. Chem. Eng. Res. Design, 62, p. 275 (1984).

33. Sheikholeslami, R., Calcium sulfate fouling- precipitation or particulate: a proposed composite model. Heat Transfer Engineering, 21, p. 24 (2000).

34. Turner, C., Klimas, S. J., Deposition of magnetite particles from flowing suspensions under flow boiling and single phase

forced convective heat transfer. Can. J. Chem. Eng., 78, p. 1065 (2000).

35. Turner, C. W., Lister, D. H., A study of the deposition of slit onto the surface of type 304 stainless steel. Can. J. Chem. Eng., 69, p. 203 (1991).

36. Vasak, F., Bowen, B. D., Chen, C. Y., Kastanek, F., Epstein, N., Fine particle deposition in laminar and turbulent flows. Can. J. Chem. Eng., 73, p. 785 (1995).

37. Walker, P., Sheikholeslami, R., Assessment of the effect of velocity and residence time in $CaSO_4$ precipitating flow reaction. Chem. Eng. Sci., 58, p. 3807 (2003).

38. Watkinson, P., Particulate fouling of sensible heat exchanger. PhD Thesis, University of British Columbia (1969).

39. Watkinson, P., Epstein, N., Particulate fouling of sensible heat exchangers. Presented at the Fourth International Heat Transfer Conference, Versailles, France (1970).

40. Whitmore, P. J., Meisen, A., Estimation of thermo- and diffusiophoretic particle deposition. Can. J. Chem. Eng., 55, p. 279 (1977).

41. Williamson, R., Newson, I., Bott, T. R., The deposition of hematite particles from flowing water. Can. J. Chem. Eng., 66, p. 51 (1988).

42. Yiantsios, S. G., Karabelas, A. J., Deposition of micron-sized particles on flat surfaces: Effects of hydrodynamic and physicochemical conditions on particle attachment efficiency. Chem. Eng. Sci., 58, p. 3105 (2003).

Chapter 3

Effect of Integrated Reservoir Tillage for in-situ Rainwater Harvesting and Other Tillage Practices on Soil Physical Properties

Haytham M. Salem[a, b], Constantino Valero[a], Miguel Ángel Muñoz[a], and María Gil-Rodríguez[a]

[a]Department of Rural Engineering, Polytechnic University of Madrid, Ciudad Universitaria s/n, 28040 Madrid, Spain
[b]Department of Soil and Water Conservation, Desert Research Center, 11753 Cairo, Egypt

ABSTRACT

There is a need for in-situ soil moisture conservation in arid and semi-arid regions due to insufficient rainfall for agriculture. For this purpose, a combination implement [integrated reservoir tillage system (RT)] comprised of a single-row chisel plow, single-row spike tooth harrow, modified seeder, and spiked roller was developed and compared to the popular tillage practices, viz., minimum tillage (MT) and conventional tillage (CT) in an arid Mediterranean environment in Egypt. The different tillage practices were conducted at tillage depths of 15, 20, and 25 cm and forward speeds of 0.69, 1, 1.25, and 1.53 m s⁻¹. Some soil physical properties, runoff, soil loss, water harvesting efficiency and yield of wheat were evaluated. The different tillage practices caused significant differences in soil physical properties as the RT increased soil infiltration, producing a rate of 48% and 65% higher than that obtained in MT and CT, respectively. The lowest values of runoff and soil loss were recorded under RT as 4.91 mm and 0.65 t ha⁻¹, whereas the highest values were recorded under CT as 11.36 mm and 1.66 t ha⁻¹, respectively. In conclusion, the RT enhanced the infiltration rate, increased water harvesting efficiency, reduced runoff and achieved the highest yield of wheat. The best tillage operating parameters appeared to be at a tillage depth of 20 cm and speed between 1.00 and 1.25 m s⁻¹.

INTRODUCTION

Water scarcity in arid and semi-arid regions due to low rainfall and uneven distribution throughout the season makes rainfed agriculture in such areas a precarious enterprise. In recent decades, there has been increased interest in the evaluation of traditional water management techniques (Prinz and Wolfer, 1999), such as rainwater harvesting for drylands agriculture, which aims to ease future water scarcity in many arid and semi-arid regions of the world.

Of the various methods of rainwater harvesting, "in-situ" systems are the simplest and cheapest approaches that can be practiced in many farming systems including those in arid and semi-arid regions. Also known as soil and water conservation systems, in-situ systems increase the amount of water stored in the soil profile by trapping or holding the rainwater where it falls (Stott et al.,

2001), which eliminates the separation between the collection and storage areas. The in-situ systems may be close to micro-catchment techniques, but they provide an alternative in arid and semi-arid regions, where precipitation is low or infrequent during the dry season. Additionally, there is a need to store the maximum amount of rainwater during the wet season for use at a later time, especially for agricultural and domestic water supply (OAS, 1997). The common in-situ rainwater harvesting techniques in arid and semi-arid regions are mulching, deep tillage, contour farming and ridging (Hatibu and Mahoo, 1999).

Soils in the Mediterranean region typically have low organic matter content, which often entails weak structure. For this reason, conventional intensive tillage systems for rainfed crops often lead to soil quality deterioration (Hernanz et al., 2002). This increases the soil's potential for erosion and also induces carbon loss, which weakens the soil's production capacity and stability. These concerns gave rise to the invention of conservation tillage practices that improve physical and biological soil properties.

Conservation tillage has several positive effects on water productivity (Rockstrom et al., 2001) compared to traditional soil and water conservation systems. Besides enhancing infiltration and soil moisture storage (Kahlon et al., 2013 and Liu et al., 2013), it reduces runoff which is then available for plant uptake during dry periods. The main limitation in stabilizing grain yields in rainfed farming systems is crop water stress caused by inefficient use of total available seasonal rainwater (McHugh et al., 2007). Therefore, technologies that use rainwater more efficiently are needed.

An alternative method to in-situ rainwater harvesting and conservation tillage is reservoir tillage, which has been defined as a system in which numerous small surface depressions are formed to collect and hold water during rainfall or irrigation to prevent surface runoff (Hackwell et al., 1991, Rochester et al., 1994,Patrick et al., 2007 and Salem et al., 2014). This approach was developed under the consideration that tillage can provide increased levels of surface storage and may represent one of the most effective means of controlling both runoff and soil erosion. A large body of research has been conducted on reservoir tillage with variations in equipment and terminology including: basin tillage, micro-basin tillage, furrows diking, furrow blocking, soil pitting, and tied-ridging (Hackwell et al., 1991, Wiyo et al., 2000, Brhane et al., 2006 and Nuti et al., 2009).

This study used Egypt as the focus region because it lies in the heart of the water scarcity problem. Egypt's rainfed agriculture is mainly concentrated in the north-western coastal zone, which extends approximately 500 km from the western city of El-Saloum on the border with Libya, to Alexandria in the east. It is bounded by the Mediterranean Sea on the north and the Sahara Desert, about 60 km on the south.

Water is particularly important in this region, as it is inhabited by an indigenous Bedouin population, 85% of which lives off of an extensive dryland production system, where barley and wheat are the main crops. Human settlements and land use are entirely dependent on rainfall and on various forms of water harvesting (Mamdouh, 1999) to increase the efficiency of runoff water use for human and animal consumption and cultivation and to minimize soil erosion.

The area's geography and hydrology are ideal for effective use of water harvesting systems. In this region, soil water management techniques must retain the maximum possible rainfall by methods that reduce storm-water runoff, improve infiltration and boost the water storage capacity of the soil. The system must also be cost effective for acceptance by the farmers.

Currently, most farmers in the northwestern coastal zone of Egypt still utilize old-fashioned crop production systems. Some farmers have switched to mechanization systems, but these systems have resulted in several problems. Some of the most common problems include added costs related to buying or renting agricultural machines, difficulties in using and maintaining these machines, crumbling of cultivated land, and the necessity for multiple machines to fulfill all agricultural processes. This last problem has been the most decisive, as farmers have often been unable to obtain all of the necessary machinery. Additionally, some farmers who own tractors as sources of power still even broadcast seeds because the region only receives rainfall during a short period of time, and so the farmers must utilize the little moisture available before it dries out.

These problems indicate a need to design integrated technologies to increase agricultural water use efficiency through rainwater harvesting while conserving the soil in rainfed areas. Researchers in this region recognized the need to develop an alternative system that was energy, water and labor efficient that could also help sustain soil and environmental quality and produce more at a lower cost. Specifically, there has been a need to produce a combination implement (integrated reservoir tillage system) to simultaneously perform multiple processes including tilling and planting in order to decrease the number of machines traveling on soil surface, which mitigates soil compaction problems (El-Saied, 2000 and Rohit and Hifjur, 2006), and consequently increases crop yields, and lowers the total cost for mechanization processes by decreasing fuel consumption, labor, maintenance cost and the cost of owning or renting machines (Tuhtaku-Ziev and Utepbergenov, 2002 and D'aene et al., 2008).

The long-term effects of conservation tillage have been well documented; however less information is available regarding the immediate effects, particularly when switching to conservation tillage from conventional tillage in such soil conditions, limited crop root development due to compaction and poor water infiltration are the major initial obstacles (Chen et al., 2005). The long-term benefit from conservation tillage cannot be achieved easily, unless producers see that the system works in a short term (Chen et al., 2005). This is a very important topic from an agronomic point of view where the adoption of conservation tillage particularly no-tillage has led to difficulties in soil workability, forcing farmers to switch to other systems (López-Garrido et al., 2014). In these cases it would be desirable that farmers initially opt for other alternatives of conservation tillage that are different from no-tillage, such as reservoir tillage (Salem et al., 2015). There is limited documentation on the immediate effects of reservoir and minimum tillage practices compared to conventional tillage on soil conditions in the north-western coastal zone of Egypt. In this region farmers frequently only consider traditional tillage with soil inversion to avoid compaction and eliminate weeds. However, less aggressive tillage practices, such as reservoir tillage and minimum tillage, could solve the problem and increase agricultural water use efficiency through rainwater harvesting without losing the advantages of conservation agriculture. Therefore, the objectives of this work were: (i) to develop and manufacture a combination implement suitable for conserving rainwater in-situ within the root-zone using a reservoir tillage tool and mechanical seeding; (ii) to increase soil moisture storage, reduce runoff, and improve infiltration of harvested water through reservoir tillage; (iii) to optimize various operating parameters that affect the performance of tillage practices; and (iv) to compare the influence of the combination implement and other popular tillage practices on soil physical properties, water harvesting efficiency, and yield of wheat.

MATERIALS AND METHODS

Site Description and Meteorological Conditions

Field experiments were carried out in Wadi Madwar located at the El-Qasr region, which lies approximately 10 km Southwest of the Marsa Matruh city and 3 km from the Mediterranean sea in Egypt's northwestern coastal zone (latitude: 31°21 08 N, longitude: 28°08 40 E, and an altitude of 30 m above sea level). The location of the study area is presented in Fig. 1. The soils of Wadi Madwar are mainly sandy loam in texture, and the average slope is between 4 and 6% in South–North direction. The climatic conditions from the Marsa Matruh meteorological station (latitude 31°20 N, longitude: 27°13 E, and an altitude of 28 m above sea level) were used to determine the meteorological data of the study area. The arid Mediterranean climatic conditions are characterized as short rainy seasons during October–March; about 85% of the total annual rainfall is recorded between December and February. During the growing season of wheat (2012–2013), the average temperature, relative humidity and total precipitation were 15 °C, 64.2% and 161.2 mm, respectively.

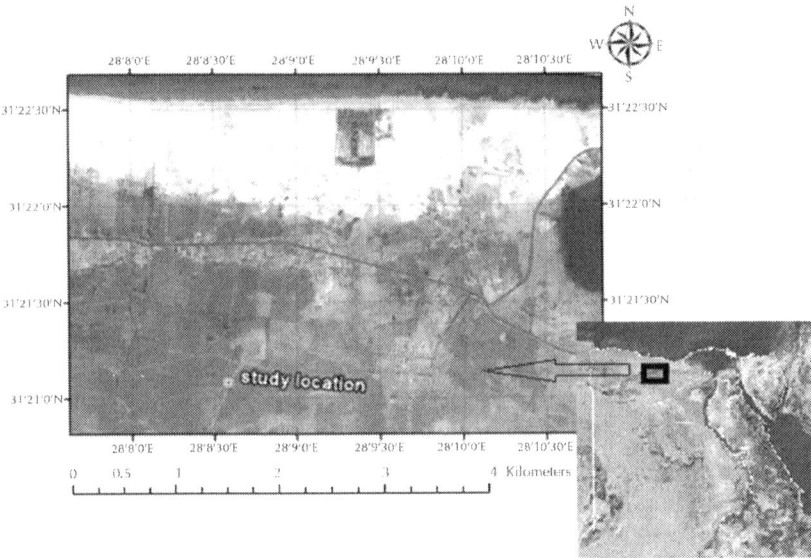

Figure 1: Location of the study area.

Combination Implement [Integrated Reservoir Tillage System (RT)]

The combination implement (Fig. 2 and Fig. 3), used in this study was manufactured from local materials to overcome the problems associated to the imported machines like cost and power requirements. The RT was mounted via three-point hitches to a tractor, and its weight was 495 kg without load (seeds). It ran on two ground removable wheels of 60 cm diameter. The main frame was constructed from rectangular iron sheet steel and had the dimensions of 140 × 165 cm. The main structure consisted of the parts defined in the following sections.

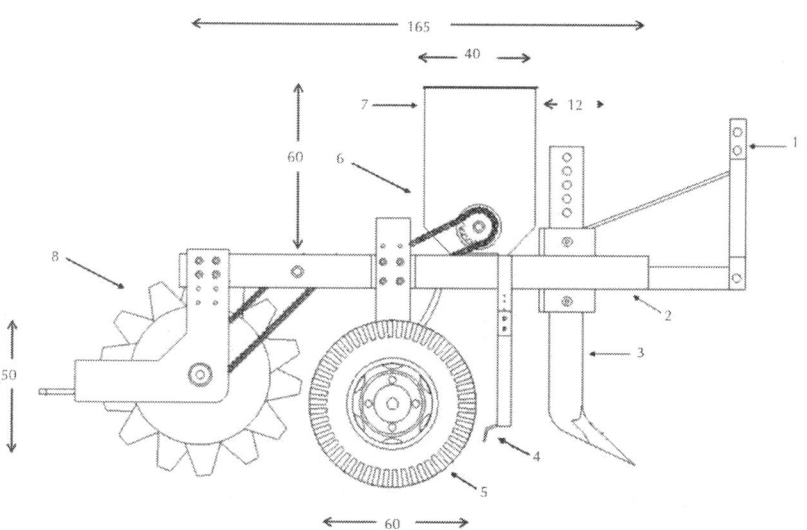

Figure 2: Side view of the combination implement (integrated reservoir tillage system), (1) upper hitch point; (2) main frame; (3) chisel plow; (4) spike-tooth harrow; (5) ground wheel; (6) feeding mechanism; (7) seed hopper; (8) spiked roller. Dimensions in centimeters.

Figure 3: (a) the main structure of the combination implement; (b) the combination implement during carrying out experiments; (c) the depressions or mini-reservoirs creating by the combination implement.

Chisel Plow

A chisel plow was used as the primary tillage tool to plow the soil for seedbed preparation. It consisted of four tines arranged in one row and mounted on straight shanks constructed from iron steel (cross section 4 × 12 cm). This shape is especially used when there is either low or heavy rainfall and can enhance the soil ability to hold water for a longer period. The distance between the shanks was 30 cm. Each shank had different levels of holes to provide different tillage depths and was mounted on the implement's main frame using stainless steel hex bolts.

Spike-tooth Harrow

A spike-tooth harrow was used as a secondary tillage tool, which had solid tines manufactured from iron that was arranged in one row and slanted to the rear to prevent clogging by debris. The harrow was fixed to plow the soil at a depth of 10 cm. Using the spike-tooth harrow in loosened soil effectively shatters and sorts soil clods and brings larger clods and aggregates to the surface. This process is recommended because of its stability under rainfall, which helps reduce soil erosion and produces a homogeneous seedbed for sowing machines.

Seeding Unit (Seed Drill)

The constructed seeding unit consisted of a grain box, a feed set, seed delivery tubes and a transmission gear attached to the machine frame for the feed set. The seed drill was used to carry out mechanical seeding in uniform rows (7 rows, 21.6 cm between each) at a controlled depth and in specified amounts. The seed box was designed and constructed from a one-millimeter thick steel sheet with dimensions of 40 × 120 × 60 cm. seven adjustable dampers were attached to the seed box. The seed metering unit consisted of a tooth roller around its circumference with a curved plate. The bottom plate regulated the slot of each feeding set so the required amount of grains could flow easily without gathering. These seeds dropped into a hinged trough that was connected to the seed box. The same trough could be adjusted to drill the seeds at different required seeding rates. The feeding unit was connected to a spiked roller through a reduction and transmission gears to be able to use different speeds for the feeding system, when the roller rotates by friction with the soil, the gears transmit the rotating movement to the feeding unit.

Spiked Roller (Reservoir Tillage Tool)

A spiked roller was developed and manufactured out of 2 mm structural steel. The tool consisted of a roller (cylindrical core has an outside diameter of 50 cm, and a width of 120 cm) that was made of 42 teeth with truncated square pyramid shapes that were welded radially on the roller to form six heptagram rings. The tooth length was 12 cm, and the spacing between the rings was approximately 9.6 cm. The radial basal spacing of the teeth was about 9.4 cm. The depressions were formed through soil compression and shearing. The gap around the base of each tooth facilitated these two interacting processes with minimal soil disturbance and compaction and improved soil penetration while providing a dam between adjacent depressions for safe storage of collected rain until infiltration.

The roller was mounted on the main implement's frame with two L shape frames on each side by stainless steel hex bolts. The roller axle (full-length, cold-rolled steel, 5 cm in diameter), Axle bearings (Pillow block type with cast steel housing and double tapered roller bearings). The spiked roller had the ability to create small depressions in the loose soil surface, which acted as reservoirs for rainwater. Each depression had a water capacity of 1 L. The final volume of these reservoirs was dependent upon the soil conditions, the expected rainfall intensity and duration, the roller's weight which can be adjusted by filling the roller to different levels with water through an opening on the bottom of the roller, and the speed of the tractor.

The spiked roller was designed to fit the modified seeder. Depressions or mini-reservoirs were made in continuous rows on both sides of the seed lines for in-situ rainwater harvesting after seeding.

Tillage Experiments and Crop Management

The field experiment was carried out on wheat during the winter season of 2012–2013. The experiment was performed in a split–split plot design with three replications. Tillage practices were maintained in the main plots, tillage depths as the sub-plot and tillage forward speeds as the sub–sub plot. The three tillage practices used in this study were (1) minimum tillage (MT), a one-pass tillage operation that uses the chisel plow and the spike-tooth harrow followed by manual spreading of seeds, (2) combination implement (integrated reservoir tillage system (RT)), and (3) conventional tillage (CT), corresponding to the traditional method (control) used by local farmers in this region of two passes of tillage operation using a chisel plow, seven tines with working width of 175 cm, followed by manual spreading of seeds after tillage. The experiment was conducted at three different levels of tillage depths (15, 20, and 25 cm) and four forward speeds of tillage operations (0.69, 1.00, 1.25 and 1.53 m s^{-1}).

The main plots occupied about 0.45 ha, each, and the main plots were divided into three subplots (about 0.15 ha) then they were divided and randomly assigned into four sub–sub plots. Each plot was 2.5 m in width and 40 m in length, with a buffer zone of 0.5 m between plots. A small area of approximately 10 m long by 2.5 m wide before the beginning of each tested plot was used to enable the tractor and implements to reach the required tillage depth and operation speed. A tractor model Ursus C-385 (63.4 kW) was used in all the experiments.

The sowing rate used in the seeder and the manual spreading was 140 kg ha^{-1} of winter wheat (Triticum aestivum L.), Giza 168 variety. Wheat was sowed at a depth of 4 cm and a seed rate of 300 seed m^{-2}, with a row spacing of 21.6 cm on 16 December 2012.

All plots were fertilized according to the common practice. A basal and single dose of superphosphate (15.5% P_2O_5) was applied at a rate of 70 kg ha^{-1}. Nitrogen fertilizer was applied as ammonium nitrate (33.5% N) at a rate of 100 kg ha^{-1} in two equal doses; the first was applied at the time of seeding, and the second was applied three weeks after seeding.

Measurements

Soil Physical Properties

Soil samples were taken with a cylindrical core at depths of 10, 20, and 30 cm and each was replicated three times per sub–sub plot. The core samples were immediately weighed and then dried at 105 °C for 24 h. Soil bulk density was measured according to Blake (1965). Further, to analyze soil physical and chemical properties, different soil samples were air-dried and sieved through a 2 mm sieve. The following analyses were conducted: The particle size distribution was determined using the pipette method according to Piper (1950). The water extract components were determined in the soil paste extract using the standard methods of analysis presented by Jackson (1969). Soil reaction (pH) was determined in the soil paste, according to the method proposed by Richards (1954). Collin's calcimeter was used for $CaCO_3$ determination according to the method described by Wright (1939). Soil organic matter content was determined following the modified Walkley and Black method (Jackson, 1969). The soil physical and chemical characteristics are shown in Table 1.

Table 1: Physical and chemical properties of the soil measured at different layers before tillage operations

Depth (cm)	pH	EC_e (dS m^{-1})	O.M (%)	$CaCO_3$ (%)	Particle size distribution %			Bulk density (Mg m^{-3})	Cone index (MPa)	Field capacity (vol. %)	Permanent wilting point (vol. %)	TAW[a](mm/20 cm)	Infiltration rate (cm h^{-1})
					Sand	Silt	Clay						
0–20	7.86	1.13	0.57	15.34	64.1	21.2	14.7	1.51	1.33	19.56	12.57	13.98	1.12
20–40	7.78	0.96	0.51	15.48	62.9	20.1	17.0	1.53	1.45	22.57	13.31	18.52	
40–60	7.82	1.11	0.34	15.79	61.6	21.2	17.2	1.54	1.49	22.05	12.54	19.02	

[a](TAW) total available water.

The infiltration rate was measured by using a double ring infiltrometer with an inner ring of 30 cm diameter and an outer ring of 60 cm diameter. The double ring was hammered into the soil to 15 cm depth. Penetration resistance (cone index) was measured by using a pentrometer model (SR-2, DIK-500). The bulk density, infiltration rate and cone index were determined before tillage and three weeks after emergence and each was replicated three times per sub–sub plot.

Runoff and Soil Loss

The runoff volume obtained on the cultivated area was collected in the receiving containers that were installed in auger holes at the down-slope edge of each sub–sub plot. Containers were covered so that rain could not enter, and evaporation was assumed to be negligible. The containers were emptied after each measurement to ensure enough space for the next storm runoff. The runoff volume was determined from the measured depth of water in each container, and the runoff coefficient was computed as the percentage of rainfall that became runoff.

Soil loss (sediment) was deposited, separated from the water, dried in an oven to a constant weight at 105 °C for 24 h, and weighed. Soil loss rate was defined by dividing sediment weight per unit area. Runoff and soil loss were measured once at the down-slope edge of each sub–sub plot.

Moisture Storage and Water Harvesting Efficiency

The field capacity and permanent wilting point moisture content were determined using undisturbed core samples collected using core sampler (19 cm length and 5.5 cm diameter) at the depths of 0–20, 20–40, and 40–60 cm. Measurements were performed once before the treatments and three times (6, 40, and 63 days after the last rain) in the dry season between 10 March and 12 May 2013 and each was replicated three times per sub–sub plot.

Soil samples were saturated for 24 h, and a pressure of 1/3 bar (for field capacity) and 15 bar (for permanent wilting point) were exerted until no further change in sample mass was observed. Based on the methodology described by Or and Wraith (2000), and IAEA (International Atomic Energy Agency, 2008), the total available water (TAW) for plant use in the root zone was computed as the difference between the moisture content at field capacity and the permanent wilting point employing Eq. (1) (Allen et al., 1998) as:

$$TAW = \sum_{i}^{n} (\theta_{FCi} - \theta_{PWPi}) Z_{ri}$$

(1)

Where TAW = total available soil water (mm); ϑ_{FCi} = moisture content at field capacity on volume base in the ith layer of the soil (%); ϑ_{PWPi} = moisture content at permanent wilting point on volume base in the ith layer of the soil (%); Z_{ri} = depth of the ith soil layer within the root zone (mm); and n = number of soil layers in the soil root zone.

The water harvesting efficiency (WHE) for one season was assessed as the ratio of the depth of stored water (S_e) in mm at the end of the rainy season to the total seasonal precipitation (P_g) in mm (Oweis and Taimeh, 1996, Boers, 1997 and Gammoh, 2013), where:

$$WHE = \left(\frac{S_e}{P_g}\right) \times 100$$

(2)

Precipitation use Efficiency and Harvest Index

Precipitation use efficiency (PUE_g) for the cropping season was calculated using the following equation (Hensley et al., 2000):

$$PUE_g = \frac{GY}{P_g + (\theta_p - \theta_h)}$$

(3)

Where PUE_g was calculated based on the grain yield (kg ha^{-1} mm^{-1}), GY = grain yield (kg ha^{-1}); P_g = precipitation during the cropping season (mm); ϑ_p = water content of the root zone at planting (mm); and ϑ_h = water content of the root zone at harvesting (mm).

The harvesting date was 12 May 2013, and the grain and biomass yield were determined from 1 m² middle area of each sub-sub plot with three replications by clipping the plants at the soil surface at harvest time. The harvest index was computed as percentage grain in the total aboveground plant biomass.

Statistical Analysis

Data were analyzed by the General Linear Model of ANOVA. The SAS (SAS/STAT, 1999–2001SAS/STAT, 1999–2001) procedure was used to test the significant differences, and Tukey's HSD (honestly significant difference) test was performed for post-hoc comparisons between treatments mean at the 95% probability level (p < 0.05). Levene and Kolmogorov–Smirnov test was applied ahead of analysis to check the normality, to ensure that assumptions of the model were met.

RESULTS AND DISCUSSION

Soil Physical Properties

Tillage practices, tillage depths, and tillage forward speeds had significant individual effects on bulk density, cone index and infiltration rate for all soil layers (0–30 cm), as shown in Table 2. This trend may be related to the maximum effective working depth among the tillage practices (15–25 cm). Simultaneously, the interaction effects between tillage practices, tillage depths and tillage forward speeds were not significant for bulk density, cone index and infiltration rate for all soil layers (0–30 cm).

Table 2: F values derived from ANOVA for bulk density, cone index, and infiltration rate under different tillage practices, tillage depths, and tillage forward speeds

Property	Source of variance	d.f.	Soil depth layers (cm)		
			0–10	10–20	20–30
Bulk density	Tillage practices (Tp)	2	8.06*	4.86*	13.09**
	Tillage depths (Td)	2	29.77**	3.41*	24.91**
	Tillage speeds (Ts)	3	26.49**	16.45**	13.36**
	Tp × Td	4	1.63	2.73*	5.61*
	Tp × Ts	6	0.54	0.16	0.75
	Td × Ts	6	0.21	0.67	1.16
	Tp × Td × Ts	12	0.32	0.17	0.39
Cone index	Tillage practices (Tp)	2	14.86**	3.19*	86.66**
	Tillage depths (Td)	2	18.55**	12.22**	131.82**
	Tillage speeds (Ts)	3	62.15**	119.85**	30.63**
	Tp × Td	4	7.78**	13.00**	4.74*
	Tp × Ts	6	3.86*	1.74	1.16
	Td × Ts	6	0.41	0.58	2.42*
	Tp × Td × Ts	12	1.51	1.29	0.91
Property	Source of variance	d.f.			
infiltration rate	Tillage practices (Tp)	2	260.12**		
	Tillage depths (Td)	2	3.14*		
	Tillage speeds (Ts)	3	23.10**		
	Tp × Td	4	12.15**		
	Tp × Ts	6	3.10*		
	Td × Ts	6	0.77		
	Tp × Td × Ts	12	0.84		

Note: * and ** indicate significant effects at 0.05 and 0.01 levels of probability, respectively.

Soil Bulk Density

Soil bulk density is a very important parameter that reflects the status of soil compaction and soil porosity. Table 3 shows the mean values of bulk density in different soil layers under different tillage practices, tillage depths, and tillage forward speeds. Bulk density generally increased with depth and was significantly affected by tillage treatments at all sampling depths in comparison with a soil without any treatments (Table 1).

There were significant differences between RT and CT in soil layers 10–20, and 20–30 cm, and the effect was not significant between MT and RT in the 20–30 cm soil layer (Table 3). Overall, in the 20–30 cm soil layer, tillage practice affected bulk density in the order: MT > RT > CT. On the other hand, in the 0–10 cm soil layer, the order changed to CT > RT > MT. In cases of MT and RT, this impact on bulk density can be attributed to the use of a spike-tooth harrow, which caused a breakdown of soil structure in the upper layer. CT was carried out in two passes, but the compaction increased due to the wheel traffic.

Soil bulk density changed not only because of constructional properties of soil tillage implements, but also because of their operational variables (Taniguchi et al., 1999 and Albiero et al., 2011). Tillage depth significantly affected soil bulk density in all soil layers except the 10–20 cm soil layer. In general, higher forward speeds caused increases in soil bulk density at all soil layers due to the production of fewer breakdowns of soil aggregates, while lower forward speeds caused decreases in soil bulk density except for the 0–10 cm soil layer, due to the effective working depth among the tillage practices (15–25 cm), for example, when forward speed increased from 0.69 to 1.53 m s^{-1}, soil bulk densities increased by 6.35, 6.97, and 4.44% for soil layers 0–10, 10–20, and 20–30 cm, respectively. This is attributable to the fact that at higher speeds of operation, the tractor tractive efficiency became very low leading to skidding. These results generally agree with earlier findings under varying soil conditions (Thakur et al., 1988, Rautaray et al., 1997 and Ahaneku and Ogunjirin, 2005).

Table 3: Mean values ± standard error of bulk density (Mg m^{-3}), and cone index (MPa) for different types of tillage practices, tillage depths, and tillage forward speeds in different soil layers

Treatments	Bulk density (Mg m^{-3})			Cone index (MPa)		
	Soil layers (cm)					
	0–10	10–20	20–30	0–10	10–20	20–30
Tillage practices						
MT	1.27 ± 0.01 [b]	1.34 ± 0.01 [ab]	1.40 ± 0.02 [a]	0.78 ± 0.02 [b]	0.97 ± 0.02 [a]	1.13 ± 0.01 [a]
RT	1.29 ± 0.01 [ab]	1.35 ± 0.02 [a]	1.38 ± 0.01 [a]	0.85 ± 0.01 [a]	0.94 ± 0.02 [b]	0.99 ± 0.02 [b]
CT	1.31 ± 0.01 [a]	1.31 ± 0.01 [b]	1.35 ± 0.01 [b]	0.84 ± 0.02 [a]	0.95 ± 0.02 [ab]	0.99 ± 0.02 [b]
SEM ±	0.006	0.008	0.006	0.009	0.008	0.009
HSD (p < 0.05)	0.021	0.027	0.021	0.029	0.029	0.031
Tillage depths (cm)						
15	1.26 ± 0.01 [c]	1.34 ± 0.01 [a]	1.41 ± 0.01 [a]	0.79 ± 0.02 [c]	0.95 ± 0.02 [b]	1.15 ± 0.01 [a]
20	1.29 ± 0.01 [b]	1.32 ± 0.02 [a]	1.37 ± 0.02 [b]	0.82 ± 0.01 [b]	0.92 ± 0.02 [b]	1.02 ± 0.02 [b]
25	1.32 ± 0.01 [a]	1.34 ± 0.01 [a]	1.35 ± 0.01 [c]	0.86 ± 0.02 [a]	0.98 ± 0.02 [a]	0.94 ± 0.02 [c]
SEM ±	0.006	0.008	0.006	0.009	0.008	0.009
HSD (p < 0.05)	0.021	0.027	0.021	0.029	0.029	0.031
Tillage speeds (m s^{-1})						
0.69	1.26 ± 0.01 [c]	1.29 ± 0.01 [c]	1.35 ± 0.02 [c]	0.73 ± 0.02 [d]	0.84 ± 0.01 [d]	0.98 ± 0.02 [c]

1	1.26 ± 0.02 [c]	1.32 ± 0.01 [b]	1.37 ± 0.01 [b]	0.79 ± 0.01 [c]	0.90 ± 0.01 [c]	1.01 ± 0.01 [bc]
1.25	1.30 ± 0.01 [b]	1.34 ± 0.01 [ab]	1.38 ± 0.02 [b]	0.85 ± 0.01 [b]	0.98 ± 0.01 [b]	1.04 ± 0.01 [b]
1.53	1.34 ± 0.01 [a]	1.38 ± 0.01 [a]	1.41 ± 0.01 [a]	0.92 ± 0.02 [a]	1.08 ± 0.02 [a]	1.12 ± 0.01 [a]
SEM ±	0.007	0.009	0.007	0.010	0.010	0.010
HSD (p < 0.05)	0.027	0.034	0.027	0.037	0.037	0.039

MT: Minimum tillage; RT: reservoir tillage; CT: conventional tillage. Different letters in the same column indicate significant differences (p < 0.05).

Soil Cone Index

Soil penetration resistance as measured by cone index has been used as an important indicator for soil compaction (Tessier et al., 1997) and crop root development (Chen et al., 2005). Table 3 represents the mean values of cone index in different soil layers with different tillage practices, tillage depths, and tillage forward speeds. Similar to bulk density, cone index increased with depth and was affected by tillage treatments at all sampling depths. Furthermore, increasing forward speed caused significant increases in cone index at all soil layers. The lowest value 0.73 MPa was recorded at a forward speed of 0.69 m s^{-1}. However, tillage practices affected cone indices in a different way at different depths. For example, no significant difference was noted in the upper layer between RT and CT, and their values were higher than MT. This was perhaps due to the effect of the roller weight used in the RT treatment to create depressions or mini-reservoirs on the soil surface and the effect of wheel traffic in the CT treatment because it was carried out in two passes. In the shallow layers, on the other hand, the RT and CT values were lower than the MT treatment.

The influence of increasing tillage depth on decreasing cone index value was clearly observed in the 20–30 cm soil layer, and the maximum reduction in cone index was 18.3% at a tillage depth of 25 cm. In general, there were significant differences between tillage depths and their values change in soil layers 0–10 and 20–30 cm, according to the effective working depth.

Infiltration Rate

Conservation tillage practices would be expected to increase infiltration and allow the water to flow deeper through the soil vadose zone. Tillage practices as well as machine operating parameters greatly affected infiltration rates. Table 4 demonstrates that there were significant differences among tillage practices. Overall, tillage practice affected infiltration rates in the order: RT > MT > CT. There was little difference between the MT and CT treatments, and the maximum mean infiltration rate of 10.69 cm h^{-1} was observed with RT at tillage depth 20 cm, which decreased to 5.99 cm h^{-1} with CT at the same tillage depth. In general, the RT increased infiltration rates by 47.52 and 64.30%, compared with MT and CT respectively, This can be explained by the fact that the large infiltration surface area created by the numerous depressions and the small depth of ponded water in the shallow depressions resulting from the RT treatment compared with MT and CT treatments (Mrabet, 2002).

Table 4: Mean values of infiltration rate (cm h^{-1}) for different types of tillage practices, tillage depths, and tillage forward speeds

Tillage practices	Tillage depths (cm)	Tillage forward speeds (m s^{-1})				Mean
		0.69	1	1.25	1.53	
MT	15	8.90 (±0.95)[a]	7.63 (±1.10)	6.95 (±0.67)	6.30 (±0.95)	7.45
RT		10.36 (±1.18)	11.07 (±0.75)	10.83 (±0.55)	8.73 (±0.59)	10.25
CT		7.11 (±0.77)	6.75 (±1.05)	5.05 (±0.69)	5.14 (±0.65)	6.01
Mean		8.79	8.48	7.61	6.72	
MT	20	8.47 (±0.72)	8.30 (±0.82)	7.63 (±0.90)	7.00 (±0.85)	7.85
RT		9.83 (±0.56)	11.47 (±0.57)	11.19 (±1.02)	10.25 (±0.77)	10.69
CT		6.83 (±0.65)	6.94 (±0.58)	5.56 (±1.09)	4.63 (±0.29)	5.99
Mean		8.38	8.90	8.13	7.29	
MT	25	6.20 (±0.89)	6.10 (±0.95)	5.83 (±0.99)	5.23 (±0.73)	5.84
RT		10.75 (±1.09)	10.68 (±0.86)	10.29 (±0.48)	9.41 (±0.96)	10.28
CT		7.67 (±0.34)	8.07 (±0.49)	6.38 (±0.69)	5.77 (±0.83)	6.97
Mean		8.21	8.28	7.50	6.80	

MT: Minimum tillage; RT: reservoir tillage; CT: conventional tillage.
[a]Figures in the parentheses are the standard deviations.

Significant differences in infiltration rate also occurred among tillage forward speed except between 0.69 and 1 m s^{-1}, where it was noted that increasing forward speed decreased infiltration rate. The minimum value of 6.72 cm h^{-1} was noticed under forward speed 1.53 m s^{-1} at a tillage depth of 15 cm.

Runoff and Soil Loss

The total rainfall for the growing season was 161.2 mm with eleven rainy days. The total depth of rainfall for the six storms that caused runoff was 79.8% of the total rainfall (Table 5), and rainfall intensities of the storms varied between 6.4 and 13.3 mm h^{-1}.

Table 5: Effect of rainfall storm event on runoff (mm) and soil loss (t ha^{-1}) in a soil without any treatments

Storm event No.	Rainfall			Runoff (mm)	Soil loss(t ha^{-1})
	Depth (mm)	Duration (h)	Intensity (mm h^{-1})		
1	9.1	3.5	2.6	–	–
2	10.3	1.6	6.4	0.81	0.31
3	33.6	3.1	12.0	4.6	0.74
4	11.6	1.2	9.7	1.02	0.44
5	41.3	3.5	13.3	5.3	0.91
6	6.5	2.3	2.8	–	–
7	18.2	1.7	11.4	2.6	0.55
8	5.3	2.8	1.9	–	–
9	13.6	1.9	7.2	1.9	0.51
10	5.2	1.5	3.5	–	–
11	6.5	1.6	4.1	–	–
Total	161.2			16.23	3.46

Table 5 shows the depth of runoff (mm) and soil losses (t ha^{-1}) in a soil without any treatments. Data illustrated that the total depth of runoff reached 16.23 mm, indicating that the average runoff coefficient reached 10.07%. Additionally, the total soil losses were 3.46 t ha $^{-1}$. Tillage practices would be expected to have a great influence on erosion control. Results in Table 6 and Fig. 4 show significant differences among tillage practices and their operating parameters on runoff and soil loss. Overall, tillage practices affected runoff in the order: CT > MT > RT. The runoff coefficient followed the same pattern. This can be explained by the great effect of soil surface indentations created by RT in a loose soil surface that acted as reservoirs or mini-depressions to increase the soil surface water storage with consequential soil surface water runoff and erosion reduction. Additionally, as a direct response to the produced runoff, there were significant differences among tillage practices. Soil losses were also consistently lower in RT treatment as compared to CT and MT treatments. The maximum reduction in soil losses were 60.8% by using RT treatment. This is explained by the fact that, when RT was used, rainfall collection in mini-reservoirs reduced runoff and its great potential to detach and transport soil particles.

This corroborates the results of many other studies (e.g., Rochester et al., 1994, Ventura et al., 2005, Patrick et al., 2007 and Salem et al., 2014). Detachment and transport by the concentrated flow is one of the main processes involved in soil erosion (Foster and Meyer, 1975). Significant differences in runoff and soil losses occurred among tillage depths, except between tillage depths of 20 and 25 cm. There were also no significant differences in runoff among different tillage forward speeds. Generally, runoff and soil losses grew by increasing forward speed, while the opposite was noticed with the tillage depth. The maximum reduction in runoff and soil losses of 11 and 9.2%, respectively, were observed at the tillage depth of 20 cm. On the other hand, increasing forward speed from 0.69 to 1.53 m s^{-1} increased runoff from 8.58 to 9.23 mm and soil losses 1.21 to 1.29 t ha^{-1}.

Table 6: Mean values ± standard error of runoff (mm), runoff coefficient (%), and soil loss (t ha^{-1}) for different types of tillage practices, tillage depths, and tillage forward speeds

Treatments	Runoff(mm)	Runoff coefficient (%)	Soil loss(t ha^{-1})
Tillage practices			
MT	10.20 ± 0.16 [b]	6.33 ± 0.10 [b]	1.44 ± 0.01 [b]
RT	4.91 ± 0.16 [c]	3.04 ± 0.10 [c]	0.65 ± 0.012 [c]
CT	11.36 ± 0.17 [a]	7.05 ± 0.11 [a]	1.66 ± 0.01 [a]
SEM ±	0.156	0.097	0.009
HSD (p < 0.05)	0.529	0.328	0.030
Tillage depths (cm)			
15	9.40 ± 0.45 [a]	5.83 ± 0.28 [a]	1.31 ± 0.07 [a]
20	8.47 ± 0.52 [b]	5.25 ± 0.32 [b]	1.20 ± 0.07 [b]
25	8.60 ± 0.51 [b]	5.33 ± 0.32 [b]	1.22 ± 0.08 [b]
SEM ±	0.156	0.097	0.009
HSD (p < 0.05)	0.529	0.328	0.030
Tillage speeds (m s^{-1})			
0.69	8.58 ± 0.53 [a]	5.32 ± 0.33 [a]	1.21 ± 0.09 [b]
1	8.60 ± 0.59 [a]	5.33 ± 0.37 [a]	1.22 ± 0.10 [b]
1.25	8.89 ± 0.61 [a]	5.51 ± 0.38 [a]	1.26 ± 0.09 [a]
1.53	9.23 ± 0.59 [a]	5.72 ± 0.37 [a]	1.29 ± 0.08 [a]
SEM ±	0.180	0.112	0.010
HSD (p < 0.05)	0.671	0.416	0.039

MT: Minimum tillage; RT: reservoir tillage; CT: conventional tillage. Different letters in the same column indicate significant differences (p < 0.05).

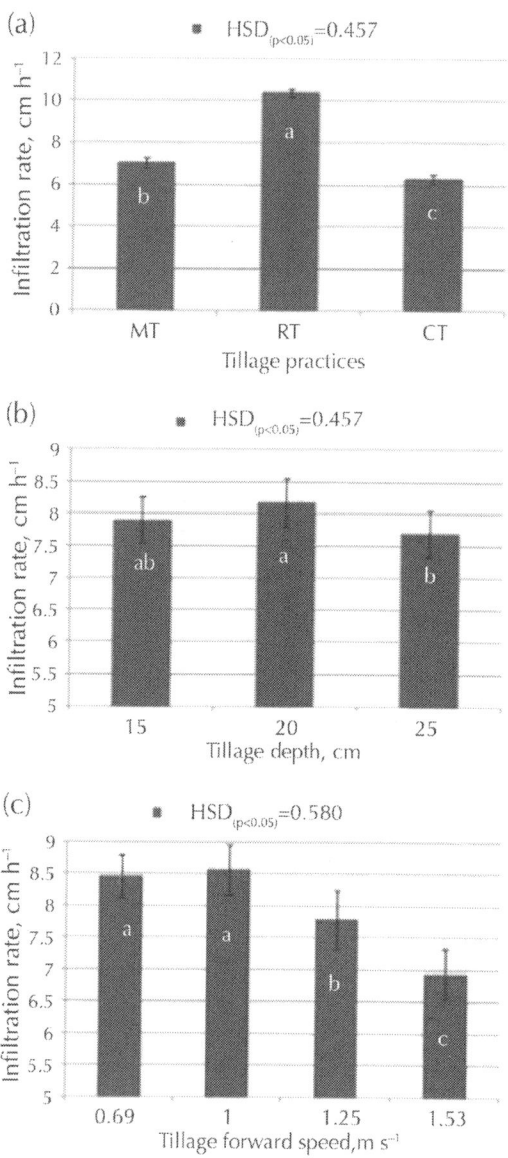

Figure 4: (a), (b) and (c) Mean infiltration rate (cm h⁻¹), under (a) tillage practices, MT: minimum tillage; RT: reservoir tillage; CT: conventional tillage, (b) tillage depths, cm, and (c) tillage forward speeds, m s⁻¹. Different letters indicate significant differences ($p < 0.05$). Line bars represent standard error.

Moisture Storage and Water Harvesting Efficiency

Summarized values of moisture content at field capacity, permanent wilting point, and the total available water (TAW) before tillage are presented in Table 1. The moisture content at field capacity varied with depth between 19.56% and 22.05% on a volume basis. The top 0–20 cm soil surface layers have a lower average moisture content at field capacity of 19.56%.

The moisture content at permanent wilting point also showed variation with depth having values between 12.54% and 13.31%. The TAW is directly related to variation in moisture content at field capacity and permanent wilting point. As a result, there was variation of TAW with depth in such a way that the highest value of TAW was found at 40–60 cm, whereas the lowest TAW was at the depth range of 0–20 cm. Variations of TAW with depth were observed as a result of the variations of moisture contents at field capacity and permanent wilting point. With tillage practice interventions, approximately 18–21 mm of unavailable water were conserved with slight reduction until the end of the growing season under all investigated techniques. Fig. 5 represents the remaining stored water as available for plant use (TAW) in different soil layers through the dry period of the growing season (t1, t2, t3, which correspond with 6, 40, and 63 days, respectively) under different tillage practices, tillage depths, and tillage forward speeds. The depth of available water under the RT treatment was significantly greater than that under MT and CT treatments in all layers, especially at (t1). Further, there were significant differences between MT and CT treatments, and the lowest TAW was observed under the CT treatment (Fig. 5a), which was clearly due to the highest infiltration rate and the more runoff being harvested from using RT. By the end of the growing season (t3), it was observed that the depth of available water under RT treatment was higher than MT and CT treatments by 44.4 and 74.5%, respectively, in all soil layers. The greatest losses were in layers 0–20, and 20–40 cm, mainly due to evaporation and plant consumption.

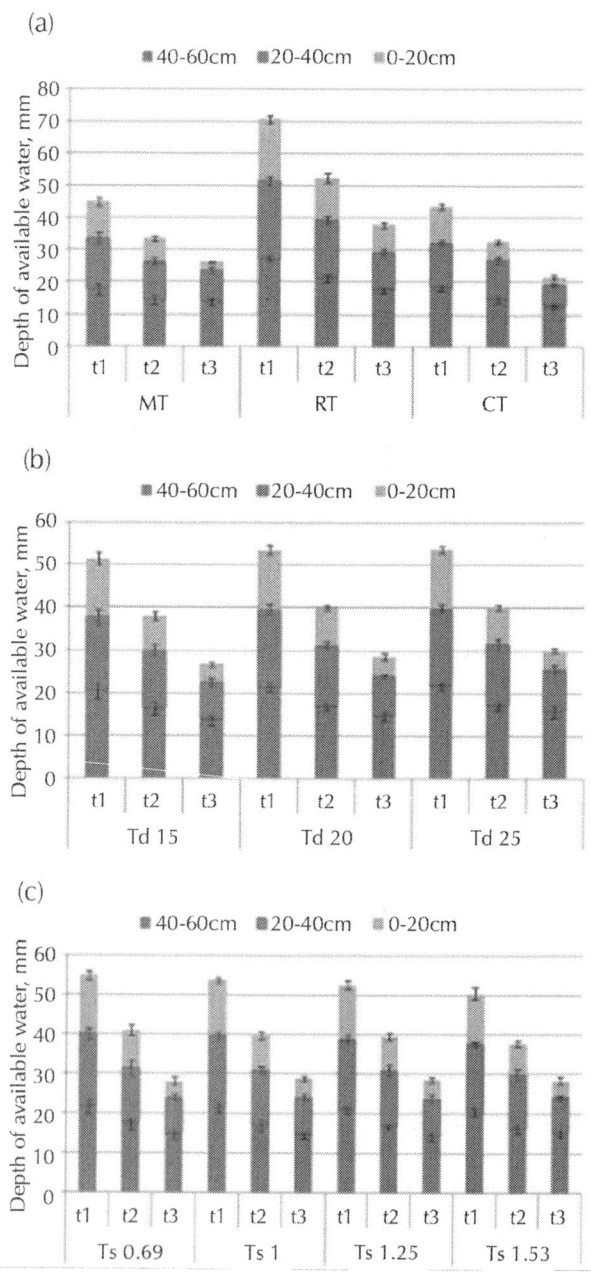

Figure 5: (a) (b) and (c) Depth of available water as distributed in the soil layers (0–20, 20–40, and 40–60 cm), through the dry period of the grow-

ing season (t1, t2, t3, which correspond with 6, 40, and 63 days, respectively, after the last rainfall event); (a) for tillage practices, MT: minimum tillage; RT: reservoir tillage; CT: conventional tillage, (b) for tillage depths (Td) 15, 20, 25 cm, and (c) for tillage forward speeds (Ts) 0.69, 1, 1.25, and 1.53 m s^{-1}. Line bars represent standard error.

The tillage operating parameters had less significant differences on TAW during each period (Fig. 5b and c), and it was observed that increasing tillage depth caused slight increases in TAW, while the opposite relationship was noticed with forward speed.

Different tillage practices had significantly different impact on the total stored water in the soil profile (S_e). The RT treatment obtained the highest value, whereas the CT treatment had the lowest value (Table 7). This can be explained by the fact that (S_e) is directly related to TAW. RT increased S_e by 39.34% more than MT and 115% more than CT. Tillage depth was not a determining factor for TAW, nor were there differences observed between the forward speeds of 0.69 and 1 m s^{-1}. The lowest values of S_e were observed under tillage depth of 15 cm and forward speed 1.53 m s^{-1}.

Table 7: Mean values ± standard error of stored water in the soil profile and water harvesting efficiency (WHE) for different types of tillage practices, tillage depths, and tillage forward speeds

Treatments	Stored water in the soil profile (mm)	WHE (%)
Tillage practices		
MT	66.59 ± 0.68 [b]	41.31 ± 0.42 [b]
RT	92.72 ± 0.71 [a]	57.52 ± 0.44 [a]
CT	43.39 ± 0.56 [c]	26.92 ± 0.35 [c]
SEM ±	0.664	0.412
HSD (p < 0.05)	2.248	1.395
Tillage depths (cm)		
15	66.39 ± 3.44 [a]	41.19 ± 2.13 [a]
20	68.05 ± 3.52 [a]	42.22 ± 2.18 [a]
25	68.26 ± 3.43 [a]	42.34 ± 2.14 [a]
SEM ±	0.664	0.412

HSD (p < 0.05)	2.248	1.395
Tillage speeds(m s^{-1})		
0.69	69.16 ± 4.40 [a]	42.90 ± 2.48 [a]
1	68.36 ± 4.18 [a]	42.40 ± 2.59 [a]
1.25	67.43 ± 4.01 [ab]	41.83 ± 2.49 [ab]
1.53	65.32 ± 3.87 [b]	40.52 ± 2.40 [b]
SEM ±	0.767	0.476
HSD (p < 0.05)	2.852	1.770

MT: Minimum tillage; RT: reservoir tillage; CT: conventional tillage. Different letters in the same column indicate significant differences (p < 0.05).

The lowest seasonal water-harvesting efficiency value (WHE) was obtained with the CT treatment, while the highest value obtained with the RT treatment (Table 7). This can be explained by the fact that WHE is directly related to S_e. Different tillage practices showed significant differences in WHE in a similar pattern to their effects on S_e.

Precipitation use Efficiency and Harvest Index

Results in Table 8 reveal that there were significant differences in grain yield among tillage practices. Overall, tillage practice affected grain yield and precipitation use efficiency in the order: RT > MT > CT. In other words, RT increased grain yield by 40.9% more than MT and 50.6% more than CT. The highest value of precipitation use efficiency of 11.87 kg ha^{-1} mm^{-1} was observed under RT, while the lowest value of 7.14 kg ha^{-1} mm^{-1} was recorded under CT treatment. In addition to previously addressed reasons, the efficiency of the RT is related to the fact that the seeding process was carried out mechanically in a controlled depth and in a specified amount of seeds per unit area using the modified seeder unit in the RT. Furthermore, it provided proper distance for optimum sunlight penetration for photosynthesis and proper depth to roots for uptake of water resulting in good water use efficiency. This corroborates the results of many other studies (e.g., Krezel and Sobkowicz, 1996 and Soomro et al., 2009).

Table 8: Mean values ± standard error of grain yield (kg ha^{-1}), precipitation use efficiency (kg ha^{-1} mm^{-1}), and harvest index (%) for different types of tillage practices, tillage depths, and tillage forward speeds

Treatments	Grain yield(kg ha^{-1})	Precipitation use efficiency(kg ha^{-1} mm^{-1})	Harvest index (%)
Tillage practices			
MT	1323.4 ± 19.75 [b]	7.78 ± 0.13 [b]	34.96 ± 0.41 [a]
RT	1864.1 ± 29.21 [a]	11.87 ± 0.20 [a]	33.03 ± 0.39 [b]
CT	1237.8 ± 21.55 [c]	7.14 ± 0.13 [c]	35.26 ± 0.56 [a]
SEM ±	22.06	0.134	0.512
HSD (p < 0.05)	74.69	0.454	1.732
Tillage depths (cm)			
15	1391.6 ± 45.55 [b]	8.35 ± 0.33 [b]	34.31 ± 0.53 [a]
20	1516.1 ± 54.53 [a]	9.22 ± 0.40 [a]	34.40 ± 0.49 [a]
25	1517.5 ± 54.31 [a]	9.23 ± 0.39 [a]	34.53 ± 0.44 [a]
SE±	22.06	0.134	0.512
HSD (p < 0.05)	74.69	0.454	1.732
Tillage speeds (m s^{-1})			
0.69	1528.3 ± 60.19 [a]	9.23 ± 0.42 [a]	34.93 ± 0.51 [a]
1	1521.2 ± 58.47 [a]	9.26 ± 0.44 [a]	35.03 ± 0.59 [a]
1.25	1453.5 ± 65.39 [ab]	8.85 ± 0.49 [ab]	33.97 ± 0.54 [a]
1.53	1397.4 ± 56 [b]	8.39 ± 0.41 [b]	33.73 ± 0.60 [a]
SEM ±	25.48	0.155	0.591
HSD (p < 0.05)	94.76	0.576	2.198

MT: Minimum tillage; RT: reservoir tillage; CT: conventional tillage. Different letters in the same column indicate significant differences (p < 0.05).

On the other hand, the seeding processes in the MT and CT treatments were carried out by manually spreading the seeds after tillage operations, which greatly decreased the germination ratio and consequently, the total yield. The biomass yield can be estimated from the harvest index value, and it shows that RT had the highest biomass yield compared with MT and CT treatments.

There were no significant differences in grain yield or precipitation use efficiency between tillage depths of 20 and 25 cm, and the lowest values were recorded under the tillage depth of 15 cm and forward speed of 1.53 m s^{-1}. In other words, it was observed that increasing tillage depth increased grain yield and precipitation use efficiency, while the opposite was noted with forward speed. For example, increasing forward speed from 0.69 to 1.53 m s^{-1} decreased grain yield and precipitation use efficiency by 8.6 and 9.1%, respectively. Tillage depth and tillage forward speed did not significantly influence the harvest index.

CONCLUSIONS

Based on the results of this research, we draw the following conclusions:

- The combination implement (integrated reservoir tillage system) enhanced infiltration rate, increased water harvesting efficiency, reduced runoff and soil losses, and exhibited the highest yield of wheat.

- The proper tillage operating parameters for the combination implement appeared to be at tillage depth of 20 cm and forward speed between 1 and 1.25 m s^{-1}. The combination implement provided, therefore, a viable option that has positive effects on soil physical properties and increased crop yields compared to minimum tillage and conventional tillage and provided an opportunity to increase agricultural water use efficiency through rainwater harvesting. Furthermore, it could be useful in saving fuel, time and production costs due to the performance of multiple processes at the same time. It is therefore desirable to encourage farmers to initially opt for this technique when switching from conventional tillage to conservation tillage. Nevertheless, continued research is needed to determine the longer term effects of these tillage practices on soil properties and crop yield.

ACKNOWLEDGMENTS

This study was financed and supported by the Desert Research Center, Cairo, Egypt, and the Rural Engineering Department, Polytechnic University of Madrid, Spain. Special thanks to workers and technicians at the Sustainable Development Center, Marsa Matruh, Egypt for their efforts as well as material support.

REFERENCES

1.   Ahaneku, I.E., Ogunjirin, O.A., 2005. Effect of tractor forward speed on sandy loam soil physical properties conditions during tillage. Niger. J. Technol. 24 (1), 51–57.

2.   Albiero, D., Maciel, A.J.S., Gamero, C.A., Lanças, K.P., Mion, R.L., Viliotti, C.A., Monteiro, L.A., 2011. Dimensional analysis of soil properties after treatment with the rotary paraplow, a new conservationist tillage tool. Span. J. Agric. Res. 9 (3), 693–701.

3.   Allen, R.G., Pereira, D., Raes, D., Smith, M., 1998. Crop evapotranspiration. Irrigation and Drainage Paper. FAO, No. 56, Rome

4.   Blake, G.R., 1965. Bulk density. In: Black, C.A. (Ed.), Methods of Soil Analysis. Part I Physical and Mineralogical Properties. SSSA Inc., Madison, WI, USA, pp. 374– 390.

5.   Boers, T.M., 1997. Rain Water Harvesting in Arid and Semi-Arid Zones (Reprinted). International Institute for Land Reclamation and Improvement, Wageningen, The Netherlands publication, no. 55, pp. 2–6

6.   Brhane, G., Wortmann, C.S., Mamo, M., Gebrekidan, H., Belay, A., 2006. Micro-basin tillage for grain sorghum

production in semiarid areas of Northern Ethiopia. Agron. J. 98, 124–128.

7.   Chen, Y., Cavers, C., Tessier, S., Monero, F., Lobb, D., 2005. Short-term tillage effects on soil cone index and plant development in a poorly drained, heavy clay soil. Soil Till. Res. 82, 161–171.

8.   D'aene, K., Vermang, J., Cornelis, W.M., Leroy, B.L.M., Schiettecatte, W., De Neve, S., Gabriels, D., Hofman, G., 2008. Reduced tillage effects on physical properties of silt loam soils growing root crops. Soil Till. Res. 99, 279–290.

9.   El-Saied, R.A., 2000. Effect of combination tillage machines on compaction soil and energy requirements. MSc. Thesis. Faculty of Agriculture, Ain Shams University p. 66

10. Foster, G.R., Meyer, L.D., 1975. Mathematical simulation of upland erosion by fundamental erosion mechanics. In: present and prospective technology for predicting sediment yields and sources. Proceedings of the Sediment-Yield Workshop, USDA Sediment Lab. Oxford, Miss. USA p. 285

11. Gammoh, I.A., 2013. An improved wide furrow micro-catchment for large-scale implementation of water-harvesting systems in arid areas. J. Arid Environ. 88, 50–56.

12. Hackwell, S.G., Rochester, E.W., Yoo, K.H., Burt, E.C., Monroe, G.E., 1991. Impact of reservoir tillage on water intake and soil erosion. Trans. Am. Soc. Agric. Eng. 34, 436–442.

13. Hatibu, N., Mahoo, H., 1999. Rainwater Harvesting Technologies for Agricultural Production: A Case for Dodoma, Tanzania. Sokoine University of agriculture, Department of Agricultural Engineering and Land Planning

14. Hensley, M., Botha, J.J., Anderson, J.J., van Staden, P.P., du Toit, A., 2000. Optimizing rainfall use efficiency for developing farmers with limited access to irrigation water. WRC Report No 878/1/00. Water Research Commission, Pretoria, South Africa

15. Hernanz, J., López, R., Navarrete, L., Sanchez-Giron, V., 2002. Long-term effects of tillage systems and rotations on

soil structural stability and organic carbon stratification in semiarid central Spain. Soil Till. Res. 66, 129–141.

16. IAEA (International Atomic Energy Agency), 2008. Field estimation of soil water content. In: A Practical Guide to Methods. Instrumentation and Sensor Technology. TCS-30 IAEA, Vienna, p. 3.

17. Jackson, A.L., 1969. Soil Chemical Analysis – Advanced Course. Pub. By Author, Dept. of Soils, Univ. of Wisc, Madison, Wisc., USA

18. Kahlon, M.S., Lal, R., Ann-Varughese, M., 2013. Twenty two years of tillage and mulching impacts on soil physical characteristics and carbon sequestration in Central Ohio. Soil Till. Res. 126, 151–158.

19. Krezel, R., Sobkowicz, P., 1996. The effect of sowing rates and methods on winter triticale grown on light soil. Roczniki Nauk Rolniczych Seria A: Produkeja Roslinna 111 (3/4), 69–78.

20. Liu, Y., Sui, Y., Gu, D., Wen, X., Chen, Y., Li, C., Liao, Y., 2013. Effects of conservation tillage on grain filling and hormonal changes in wheat under simulated rainfall conditions. Field Crops Res. 144, 43–51.

21. López-Garrido, R., Madejón, E., León-Camacho, M., Girón, I., Moreno, F., Murillo, G. M., 2014. Reduced tillage as an alternative to no-tillage under Mediterranean conditions: a case study. Soil Till. Res. 140, 40–47.

22. Mamdouh, N., 1999. Assessing Desertification and Water Harvesting in the Middle East and North Africa. Bonn, Germany: Zentrum für Entwicklungsforschung (ZEF).

23. McHugh, O.V., Steenhuis, T.S., Abebe, B., Fernandes, E.C.M., 2007. Performance of in situ rainwater conservation tillage techniques in dry spell mitigation and erosion control in the drought-prone North Wello zone of the Ethiopian highlands. Soil Till. Res. 97, 19–36.

24. Mrabet, R., 2002. Stratification of soil and organic matter under tillage systems in Africa. Soil Till. Res. 66, 119–128.

25. Nuti, R.C., Lamb, M.C., Sorensen, R.B., Truman, C.C., 2009. Agronomic and economic response to furrow diking tillage in irrigated and non-irrigated cotton (Gossypium hirsutum L.). Agric. Water Manage. 96, 1078–1084.

26. OAS, 1997. Organization of American States). Source Book of Alternative Technologies for Freshwater Augmentation in Latin America and the Caribbean. Unit of Sustainable Development and Environment General Secretariat. Organization of American States Washington, D.C.

27. Or, D., Wraith, J.M., 2000. Soil water content and water potential relationships. In: Sumner, M.E. (Ed.), Handbook of Soil Science. CRC Press, Boca Raton, FL, USA, pp. 54–64.

28. Oweis, T.Y., Taimeh, A.Y., 1996. Evaluation of a small basin water harvesting system in the arid region of Jordan. Water Resour. Manage. 10, 21–34.

29. Patrick, C., Kechavarzi, C., James, I.T., Dogherty, M.O., Godwin, R.J., 2007. Developing reservoir tillage technology for semi-arid environments. Soil Use Manage. 23 (June), 185–191.

30. Piper, C.S., 1950. Soil and plant analysis. A Monograph from Wait Agric. Research Institute, Univ. of Adelaide, Australia

31. Prinz, D., Wolfer, S., 1999. Traditional techniques of water management to cover future irrigation water demand. Z. f. Bewässerungswirtschaft 34 (1), 41–60 ISSN 0049-8602.

32. Rautaray, S.K., Watts, C.W., Dexter, A.R., 1997. Pudding effects on soil physical parameters. AMA 28 (3), 37–40.

33. Richards, L.A., 1954. Diagnosis and improvement of saline and alkali soils. U.S. Dept. of Agric, Hand Book N, pp. 60–$9.

34. Rochester, E.W., Hill, D.T., Yoo, K.H., 1994. Impact of reservoir tillage on run-off quality and quantity. Trans. Am. Soc. Agric. Eng. 37 (4), 1183–1186.

35. Rockstrom, J., Barron, J., Fox, P., 2001. Water productivity in rainfed agriculture: challenges and opportunities for smallholder farmers in drought-prone tropical agro-systems.

Paper Presented at an IWMI Workshop, Colombo, Sri Lanka, and November 12–14

36. Rohit, K.S., Hifjur, R., 2006. An approach for draft prediction of combination tillage implements in sandy clay loam soil. Soil Till. Res. 90, 145–155.

37. Salem, H.M., Valero, C., Muñoz, M.A., Rodríguez, M.G., Barreiro, P., 2014. Effect of reservoir tillage on rainwater harvesting and soil erosion control under a developed rainfall simulator. Catena 113, 353–362.

38. Salem, H.M., Valero, C., Muñoz, M.A., Rodríguez, M.G., Silva, L.L., 2015. Short-term effects of four tillage practices on soil physical properties, soil water potential, and maize yield. Geoderma 237–238, 60–70.

39. SAS/STAT, 1999–2001. SAS7. In: STAT User's Guide (Release8.2). SAS Instinct. Cary, NC.

40. Soomro, U.A., Rahman, M.U., Odhano, E.A., Gul, S., Tareen, A.Q., 2009. Effects of sowing method and seed rate on growth and yield of wheat (Triticum aestivum). World J. Agric. Sci. 5 (2), 159–162.

41. Stott, D.E., Mohtar, R.H., Steinhardt, G.C., 2001. Water conservation, harvesting and management (WCHM) – Kenyan experience.

42. Taniguchi, T., Makanga, J.T., Ohtoma, K., Kishimoto, T., 1999. Draft and soil manipulation by a moldboard plow under different forward speed and body attachments. Trans. ASAE 42, 1517–1521.

43. Tessier, S., Lachance, B., Laguë, C., Chen, Y., Chi, L., Bachand, D., 1997. Soil compaction reduction with a modified one-way disker. Soil Till. Res. 42, 63–77.

44. Thakur, T.C., Yadav, A., Varshney, B.P., Chand, P., 1988. Effects of load and speed on performance of clod crushers. AMA 19 (4), 1520.

45. Tuhtaku-Ziev, A., Utepbergenov, B.K., 2002. Combined implements for simultaneous loosening and leveling of soil surface. J. Agric. Mech. Asia Africa Latin Am. 33 (2), 15–16.

46.   Ventura, E., Norton, L.D., Ward, K., López-Bautista, M., Tapia-Naranjo, A., 2005. A new reservoir tillage system for crop production in semiarid areas. ASAE Annual Meeting. American Society of Agricultural Engineers. http://asae.frymulti.com/ request.asp?

47.   Wiyo, K.A., Kasomekera, Z.M., Feyen, J., 2000. Effect of tied-ridging on soil water status of a maize crop under Malawi conditions. Agric. Water Manage. 45, 101–125.

48.   Wright, C.H., 1939. Soil analysis. A Handbook of Physical and Chemical Methods. Thomas Murby and Co., London

Diagenetic Controlled Reservoir Quality of South Pars Gas Field, an Integrated Approach

Vahid Tavakoli, Hossain Rahimpour-Bonab, and
Behrooz Esrafili-Dizaji

Department of Geology, College of Science, University of Tehran,
6455 Tehran, Iran

ABSTRACT

The Dalan-Kangan Permo-Triassic aged carbonates were deposited
in the South Pars gas field in the Persian Gulf Basin, offshore Iran.
Based on the thin section studies from this field, pore spaces are
classified into three groups including depositional, fabric-selective
and non-fabric selective. Stable isotope studies confirm the role

of diagenesis in reservoir quality development. Integration of various data show that different diagenetic processes developed in two reservoir zones in the Kangan and Dalan formations. While dolomitisation enhanced reservoir properties in the upper K2 and lower K4 units, lower part of K2 and upper part of K4 have experienced more dissolution. Integration of RQI, porosity-permeability values and pore-throat sizes resulted from mercury intrusion tests shows detailed petrophysical behavior in reservoir zones. Though both upper K2 and lower K4 are dolomitised, in upper K2 unit non-fabric selective pores are dominant and fabric destructive dolomitisation is the main cause of high reservoir quality. In comparison, lower K4 has more fabric-selective pores that have been connected by fabric retentive to selective dolomitisation.

INTRODUCTION

Diagenetic processes can enhance, create and/or destroy porosity in carbonate rocks. In essence, several studies signify that the reservoir quality is mainly controlled by the pore-geometry, which, in turn, is mainly determined by various diagenetic processes (Baron et al., 2008, Cerepi et al., 2003, Stentoft et al., 2003 and Tucker and Bathurst, 1990). Accordingly, presence or absence of diagenetic imprints along with their type and intensity, play an important role in defining the ultimate reservoir quality and characteristics. As shown by many authors (Abid and Hesse, 2007, Alvarez and Roser, 2007, Ehrenberg et al., 2008, Elias et al., 2004, Esrafili-Dizaji and Rahimpour-Bonab, 2009, Rahimpour-Bonab, 2007 and Rahimpour-Bonab et al., 2009) variations in diagenetic alterations such as porosity, permeability and lithology, could produce zones with different reservoir properties and so different petrophysical behaviors. The petrophysical properties such as total and effective porosity, permeability, pore-throat size and distribution are substantially affected by diagenesis type and intensity. Thus, as emphasised by many authors (Rahimpour-Bonab, 2007), carbonates reservoirs which are prone to intense diagenetic alterations, could be compartmentalised (segmented) and represent variable petrophysical properties, even in small scales. So, a procedure for the identification and characterisation of comparable diagenetic units from a petrophysical point of view, would be useful to resolve some of the key challenges faced in the exploration and production of carbonate reservoirs. As a rule of thumb, in carbonate units by progress in the diagenetic processes and so overprints, pore types and throats evolve imparting changes in petrophysical properties.

This article describes such a procedure and discusses its implications in understanding diagenetic processes that affect reservoir quality in a carbonate reservoir. Based on the integration of geological, petrophysical and isotope data as well as wireline logs, main reservoir controls were identified. The effect of each factor has been discussed in detail. In this regard, the Permo-Triassic (Upper Dalan and Kangan) reservoir intervals of South Pars gas field in Iran have been investigated. Data from three wells in offshore Persian Gulf have been documented and discussed. This procedure is particularly useful for porosity evolution studies that, in turn, are important for hydrocarbon explorations.

GEOLOGICAL SETTING AND STRATIGRAPHY

The South Pars gas-condensate field, and its southern extension, the North Dome, are located in the Persian Gulf (Fig. 1). It is part of the huge NNE-SSW trending Qatar-Fars Arch. Qatar Arch is located in the interior platform of the Arabian Plate and bounded by the Zagros folded belt to the north and northeast. The South Pars field is actually the northern extension of the North field, located in the Iranian territory.

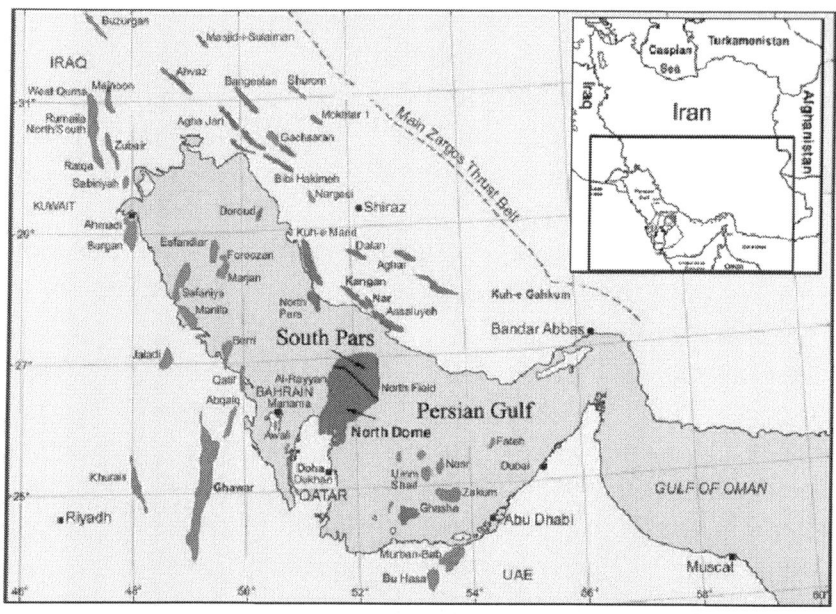

Figure 1: Geographical and geological setting of the South Pars gas field. Main hydrocarbon fields in Persian Gulf and adjacent areas are shown. The main Zagros trust belt is clear.

The major structural features of the area are the results of two main tectonic phases: the first tectonic phase is the Amar Collision that was occurred 620–640 My ago along the north-south trending Amar weak zone in the Arabian Shield. The second tectonic phase corresponds to the Najd Rift System that had occurred 530–570 million years ago, with about 300 km width and a general northwest–southeast trend, parallel to the Zagros Mountains (Al-Husseini, 2000). These major tectonic events, especially the first one, are responsible for the formation of the Qatar/Fars High and other similar structures such as the Ghawar High. Subsequent rejuvenation of the north-south structural trends controlled the structural development of the sedimentary cover and the distribution of the hydrocarbon reservoirs in the area. The Qatar-Fars High, the same as the other northerly trends, is of Precambrian basement origin. The Khuff Formation of the Arabian Plate and its equivalents Dalan and Kangan formations of Iran are thus interpreted as reflecting a major tectono-eustatic event related to the onset of rapid thermal subsidence of the early Neo-Tethys passive margin in Arabia and Iran, and the drowning of its rift shoulders (Insalaco et al., 2006).

In the Late Permian, increased accommodation space related to stretching of the crust accompanied the formation of the Neo-Tethys Ocean along the Oman-Zagros suture. The base of the resulting megasequence consists of continental to marine sandstones and shales supplied from the west. These were followed by the deposition of extensive carbonates and anhydrites (Khuff Formation in Saudi Arabia and Oman; Dalan and Kangan formations in Iran; [Al-Aswad, 1997]) over the entire Arabian shelf in shallow marine to tidal flat environments (Konert et al., 2001). Geologic traverse through Saudi Arabia to South Pars has been shown in Fig. 2.

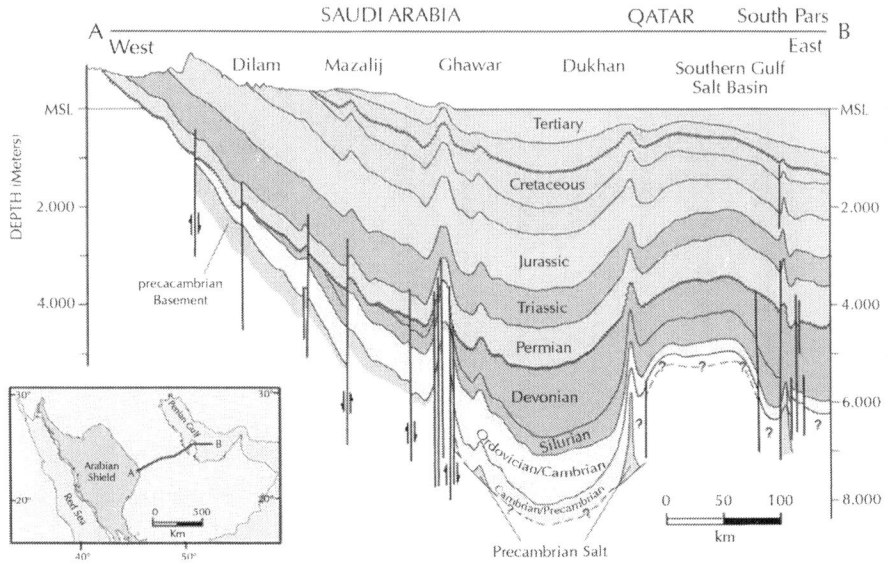

Figure 2: Geologic traverse through Saudi Arabia to Qatar and South Pars shows the correlation between formations in these regions Une coupe géologique à travers l'Arabie saoudite jusqu'au Quatar et à South Pars montre la corrélation entre les formations de ces régions.Modified from Konert et al. (2001).

The studied field was discovered in 1990 by drilling well SP-01 that encountered huge gas reservoirs in the Kangan-Dalan formations. In the South Pars field, gas accumulation is mostly limited to the Permian–Triassic stratigraphical units that became prospective during the 1970s following delineation of enormous gas reserves (Bordnave, 2008). Generalized Permian–Triassic stratigraphy of the South Pars field is shown in Fig. 3. In the South Pars field, the Early Permian–Early Triassic has been divided into Faraghan (Early Permian), Dalan (Late Permian) and Kangan (Early Triassic) formations (Kashfi, 2000). The Faraghan Formation is overlain disconformably by Dalan Formation that is subdivided into K5, Median Anhydrite (Nar member), K4 and K3 from bottom to the top, respectively (Fig. 3). The Median Anhydrite rests on the K5 dolomite. The succession is followed with the K4 unit that consists of dolomite and limestone with some anhydrite intervals. The K3 member consists of dolomite with lesser amounts of dolomitic limestone. Another main reservoir unit, named Kangan in Iranian nomenclature, conformably overlies on the Late Permian strata of K3 (uppermost Dalan Formation). Szabo and Keradpir (1978) refer Kangan Formation to Early Triassic, but recent investigations by Rahimpour-Bonab et al. (2009) indicate an important gap in the Permo-Triassic boundary in this field. Limestones and dolomites of K2 and anhydritic dolomite, dolomite and limestones of K1 are constituents of this Formation. The Kangan Formation is terminated with Dashtak Formation as an efficient cap rock. Main reservoir intervals are K2 and K4 (Aali et al., 2006, Ehrenberg, 2006, Moradpour et al., 2008, Rahimpour-Bonab, 2007 and Rahimpour-Bonab et al., 2009). The Field is a good representative of heterogeneous carbonate-evaporite reservoirs in the world. The lithological variations from limestone to dolomite and anhydrite control the reservoir properties that caused immense lateral and vertical changes in the porosity types and values. These, along with complicated diagenetic history, induced important heterogeneities in the reservoirs.

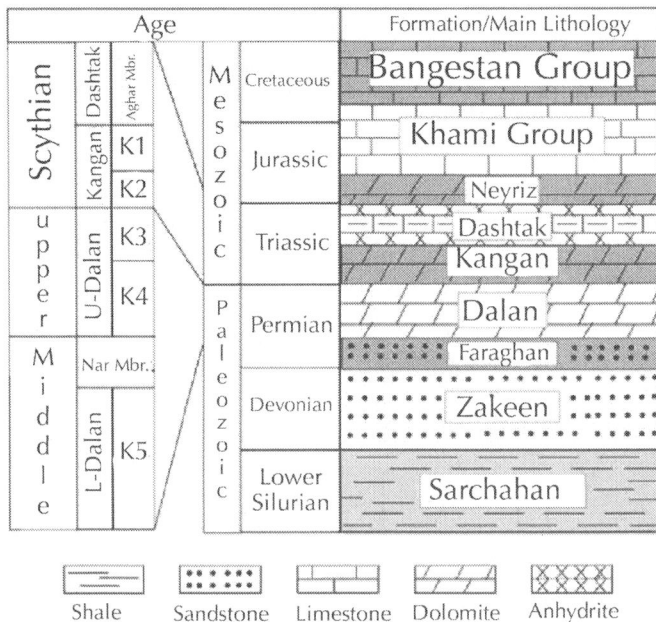

Age					Formation/Main Lithology

Figure 3: Generalized stratigraphic chart of the South Pars gas field (not to scale). Formations and main lithology from Lower Silurian to Cretaceous have been shown. The age and position of Kangan and Dalan formations in stratigraphic column are mentioned. Subdivisions of these formations are clipped.

MATERIAL AND METHODS

Our dataset consists of wireline logs, core and thin sections in about 30 cm intervals of three wells from 1200 m cores of Kangan and Dalan formations from South Pars gas fields in Persian Gulf (wells OFA_1, OFA_2 and OFA_3). Wireline logs, cores and thin sections for 12 other wells have been used for general fields study. For facies analysis, Dunham texture scheme was used together with sedimentary structures and fabrics, grain size, rock composition, and diagnostic allochems such as ooids, pelloids and shells. With careful petrographic examinations using core samples, thin sections and scanning electron microscopy (SEM), facies types, different diagenetic imprints, distribution patterns of various grains and pore types were determined. Thin sections were stained with alizarin red-S for distinguishing calcite from dolomite. Types and abundance of porosity were determined using digital point counting method. For this approach 68 thin-sectioned samples, representative of all lithofacies along the studied intervals, were examined. A total of 250 points was counted per image and between six to 10 photos prepared covering entire area of a thin section. Cleaned and dried core plug samples have been used to measure the porosity and permeability. Porosity values have been obtained by means of Boyle's law. In the routine core analysis, absolute permeabilities were obtained by gas flowing, either air or nitrogen, through the samples. Pore-throat size distributions have been achieved using mercury intrusion tests. Carbon and Oxygen isotopic data of the 348 samples from two wells were available from earlier studies (Rahimpour-Bonab et al., 2009) of the Dalan and Kangan formations. Carbon and Oxygen isotopic data from 137 limestone and dolomite samples were added in this study. Samples were powdered and then bulk powdered samples sent to the Texas A and M University's Laboratory for oxygen and carbon stable isotope analysis. For better understanding of overall distribution of porosity type, velocity deviation log (VDL) was calculated using Anselmetti and Eberli formula (Anselmetti and Eberli, 1993) and correlated with stable isotopes, core porosity and permeability, reservoir quality and diagenetic processes. All cores were studied in detail for lithofacies determination and sedimentological logs preparation.

RESULTS

Depositional Environment Summery

In the studied formations, 14 facies types were distinguished. Theses facies and their inferred depositional model are shown in Fig. 4.

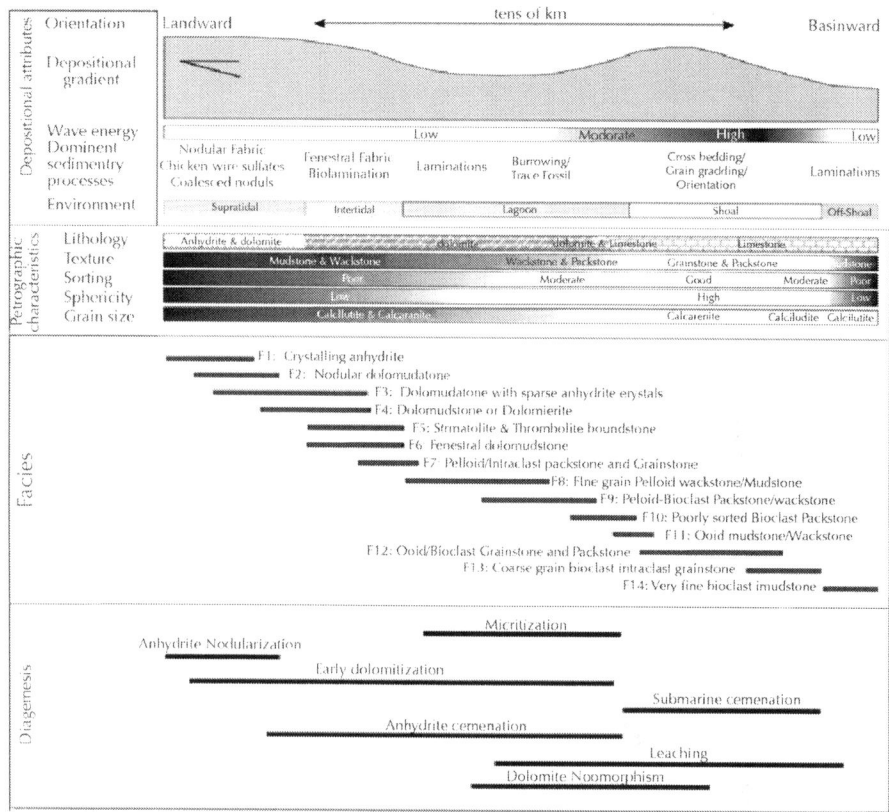

Figure 4: Depositional environment, petrographic characteristics, facies types and distribution and diagenesis processes through the studied formations. There is not a direct relationship between these facies and reservoir quality.

In general, five major facies belts can be recognised in the field scale: the supratidal facies association is characterised by anhydrite lithology and dolomudstones with layered, massive and nodular (chicken-wire) structures. The intertidal facies association consists of dolomudstones with fenestral and homogeneous fabrics and with algal mat boundstone and anhydrite crystals. The lagoon facies association consists of bioturbated pelloid dominated packstones to laminated mudstones. The shoal facies association is characterised by oolitic, bioclastic and intraclastic grainstones with flow structures and oolitic mudstones in back shoal setting. Finally, the offshoal facies association consists of very fine bioclastic and laminated mudstones. Such depositional setting for upper Dalan-Kangan strata (upper Khuff equivalent) are also recognised and documented by other workers in outcrop sections (Zagros outcrops) and other gas fields in the Persian Gulf basin (Al-Aswad, 1997, Alsharhan and Kendall, 2003 and Insalaco et al., 2006).

This depositional environment is closely comparable to the present-day Persian Gulf carbonate system, which is well-known and well documented (Alsharhan and Kendall, 2003). Seemingly, the studied strata were deposited over the inner part of such a carbonate system. Regarding to the biological evidences, inner ramp setting with a maximum water depth of 30 m is proposed here.

Petrophysical examinations of different facies types show strong variations in the porosity and permeability of the studied formations (Esrafili-Dizaji and Rahimpour-Bonab, 2009). This indicates that reservoir quality of these formations was strongly influenced by the later diagenetic processes. A comprehensive description of the upper Dalan and Kangan sedimentary environments has been shown by Esrafili-Dizaji and Rahimpour-Bonab (Esrafili-Dizaji and Rahimpour-Bonab, 2009).

Diagenesis

Based on the mineralogical compositions, cements types and occurrences, as well as main microfabrics, diagenetic environments of the studied units are distinguished and assigned to the marine, meteoric and burial realms (as observed by some authors, such as Esrafili-Dizaji and Rahimpour-Bonab, 2009,Moradpour et al., 2008 and Rahimpour-Bonab et al., 2010).

Marine Diagenesis

Generally, the early diagenetic overprints of marine origin include micritisation, marine cementation and early stages of compaction.

The petrological features of this realm include fibrous cements, skeletal micritisation, orientation of the elongated grains, early anhydrite cementation and slight deformation of grains. Marine aragonitic isopachous fringes around ooids, skeletal and non-skeletal fragments have been stabilized to calcite in the meteoric realm. These marine cements are immediately followed by clear calcite spar of phreatic or burial environment. In addition to cementation, microbial micritisation was a common process in the Kangan-upper Dalan carbonates. In the oolitic grainstones and packstones, some of the ooids and skeletal fragments show a micritic envelope, a thin and black coating around the grains. In some cases, the grain has been dissolved and the micritic envelope is still visible (Fig. S1). It is not present on all grains and its recognition can be difficult where there is isopachous cement. Circumgranular isopachous calcite cement with bladed rind is found in the oolitic grainstones either directly on the ooid or on its micritic envelope. This cement often shows dissolution either on the grain side or on the outer side.

Anhydrite cementation and nodule formation has occurred in hypersaline conditions in restricted and shallow marine environments. A comprehensive study of anhydrite formation shows that most of the anhydrites in Kangan and Dalan formations in South Pars field are primary in origin (Rahimpour-Bonab et al., 2009). This is inferred from anhydrite/dolomite relationships and cross-cutting of anhydrite with stylolites. Early dolomitization accompanied this anhydrite cements. Poikilotopic anhydrite that fills intergranular porosity is typical fabric of this diagenetic realm.

The major dolomitization phase occurred during syndepositional to shallow-burial conditions before significant burial. Petrographic evidences which support these conditions include fine crystals of dolomite that occur in early pore spaces, preservation of depositional and early diagenetic characteristics (bioturbation, micritization and marine cementation) by this type of dolomite, stylolitization and fracturing post-date dolomitization and anhydrite fabrics associated with replacive dolomites which are cut by stylolites. Most syndepositional to shallow-burial dolomites in Kangan and upper Dalan formations are fabric-retentive in texture (Fig. S1b,c).

Slightly-compacted contacts between dolomitized grains suggest that this dolomitization took place before significant burial.

Meteoric Diagenesis

The diagenesis stages continue with meteoric diagenesis during which metastable skeletal and non-skeletal grains were dissolved, generating secondary fabric- and non-fabric selective porosity. Typical freshwater cementations and dissolutions along with the polymorphic transformations of primary marine minerals all are an indication of the meteoric diagenetic realm. Petrological features typical of this realm include equant sparry calcite cement, vadose calcite silt, intragranular and intergranular dissolution pores, solution-enlarged vugs and minor gravitational cementation occurred and was recognized using optical microscopy and SEM studies. Several samples have well-developed isopachous fringes of equant calcite crystals, a few tens of μm in size (Fig. S1d). These strongly resemble cements typically formed in the shallow meteoric phreatic zone of emergent ooid shoals (e.g. Halley and Harris, 1979 and James and Choquette, 1984). Some cements progressively occluded primary pore spaces. Aggrading limestone neomorphism occurred widely in various parts of the studied units (Fig. S1e).

Common textures generated in shallow meteoric diagenetic zone of Kangan and Dalan formations include fenestrae (Fig. S1f), gas-escape structures, desiccation cracks (Fig. S1g), swirling structures and planar to curviplanar grain contacts. Rocks display evidence of physical compaction but no or poorly developed stylolites, are defined to have undergone shallow-burial diagenesis.

In the studied formations, the dissolution, which usually changes and overprints other diagenetic features, could be classified into two types, i.e. freshwater (meteoric) and burial dissolution. Meteoric dissolution which is extensive and mainly fabric-selective (Figs. S1h and S2a) has occurred in the shallow meteoric realm while sediments were not completely lithified. This fabric-selective dissolution is related to the type and nature of the allochems (Fig. S2b). In wackestones and mudstones, the dissolution voids, which are isolated and irregular, show small size and partial distribution (Fig. S2c).

Burial Diagenesis

Grading recrystallisation and chemical compaction indicate a burial diagenetic environment and its petrographic characteristics include coarse mosaic calcites with undulatory extinction, stylolites, sutured and concave-convex contacts between grains (Fig. S2d), fractures and dissolution vugs along the stylolites. Compaction, cementation and dissolution processes took place during burial of the Kangan and upper Dalan reservoir. Mechanical compaction has variably reduced porosity in oolitic grainstones and has resulted in nested fabrics (Fig. S2e). Chemical compaction, or pressure-dissolution, is an important diagenetic process in a burial environment. In addition to production of pressure-dissolution fabrics such as stylolites (Taghavi et al., 2006), the chemical compaction leads to the dissolution of grains and matrix, which is an important source of burial cements. Local and anomalous increase in horizontal permeability in some units could be related to theses stylolites (Fig. S2f).

Samples with well-developed isopachous early cements (Fig. S2g) were relatively more resistant to mechanical compaction than those without. Such samples were also less affected by pressure dissolution and contain fewer concave-convex grain contacts. Pressure dissolution also resulted in localised formation of solution seams that concentrated non-carbonate residues (such as clays and oxide minerals). Dolomite cements are volumetrically insignificant in studied intervals. Formation of replacive dolomites led to the formation of fabric-destructive (coarse and idiotopic) dolomite bodies. Most of the fabric-destructive dolomites show stylolites "ghosts" and so they probably formed during deeper burial because they post-date stylolitization. Crystals have cloudy (inclusion-rich) cores and clear (inclusion-free) rims. Euhedral to subhedral shapes and enlarged crystal sizes result from increasing burial depths and temperatures (Fig. S2h). The same dolomite neomorphism during burial has been reported from the Khuff carbonates offshore Dubai (Videtich, 1994), and east of the Qatar Arch (Alsharhan, 2006).

Late stage anhydrite and calcite cement occluded early porosity and late fractures, which are locally important. Chicken-wire anhydrite is a product of compaction of earlier-formed anhydrite textures during burial. Burial anhydrite is also present as fracture filling cement. Late stage anhydrite forms along stylolite in some places. Fracturing and saddle dolomite cementation occurred during late phases of burial diagenesis. Theses two impacts are insignificant in overall studied units. (Ehrenberg, 2006, Guadagno and Nunziata, 1993 and Kirmaci, 2008).

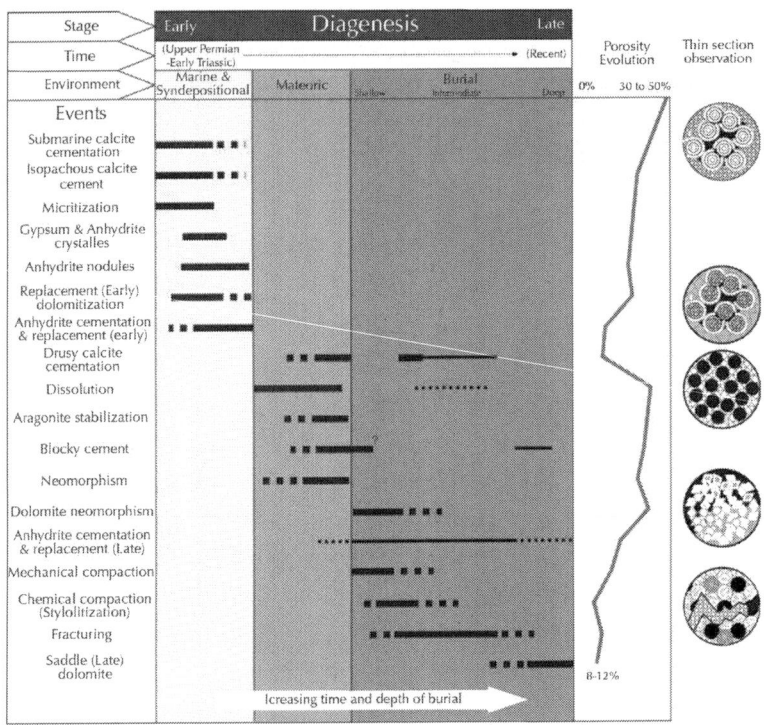

Figure 5: Generalized diagenetic sequence of the studied formations. Time, stage and environment of various diagenetic events are clear. Porosity variation in different stages is apparent. Evidences for various diagenetic environment are mentioned.

Burial dissolution voids usually occur in units with well-developed chemical compaction. Although the burial dissolution is not common in the study area, a few voids and solution-enlarged fractures along the stylolites could be observed.

Generalized diagenetic sequences of the studied formations along with the related porosity evolution are shown in Fig. 5, clarifying diagenetic events, environments and their relative time of formation. There is no considerable porosity-depth correlation in the field such as other fields in the Zagros Fold Belt (Ehrenberg et al., 2008).

Diagenetic Imprints on the Studied Units

K4 Unit: This subunit, 170 m thick, consist of limestone at the top and dolomite below. Based on core analysis this unit has the highest average porosity (15.61%) with average permeability of 36.27 md. Most of the diagenetic processes in this zone are dissolution and dolomitisation. Primary pores are present especially in the upper parts. These pores have been connected in the lower parts by dolomitisation. Planar-e dolomites are most common in this unit. Fabric-selective dolomitisation is dominant in domolitic intervals of this unit

Dissolution of both matrix and grains are obvious in the upper part of this unit. Fracturing has had a minor influence on porosity generation, in some cases, they have had an effect on the permeability increase. Fabric selective pores connected by fabric retentive dolomitisation are dominant porosity type in the lower part. Anhydrite is not common in K4 unit and when present, it could be seen as poikilotopic texture. As a result of increasing compaction, solution seams and stylolites are developed. These are not common in K4 unit.

K3 Unit: This unit is 133 m thick and consists of alternating limestones, dolomite and limy dolostones. Average porosity of this unit is 5.92% and average permeability is 19.21 md. This unit shows significant calcite and dolomite cementation and hence poor-to-fair reservoir quality. These cements were sometimes recrystallised.

Anhydrite has been developed in various parts of the unit especially as chicken-wire texture. Calcite cement is also common. Micritisation is abundant in the Permo-Triassic boundary but could be seen in the entire unit. Ragged dark rims around the skeletal grains and irregular dark holes within them are frequently observed under the microscope and by SEM studies. They have been formed either by repeated microbial boring and subsequent micrite filling or by recrystallisation of early minerals. Some grains are fully replaced by micrite, forming black micritic casts. Bioturbation could be seen below the boundary. Compaction related phenomena such as stylolite, orientation of the elongated grains, and slight deformation of grains are present. Pervasive cementation has filled most of the pores. The remaining porosities include intercrystalline and minor fractures. Planar-s dolomites have been developed in middle part of this unit.

K2 Unit: This unit with total 44.5 m thickness consists of 15.5 m of limy dolomite and 29 m of dolomitic limestone. With average porosity of 9.71% and average permeability of 38.81 md, this unit has a good reservoir quality. The abundance of anhydrite cement (filling interparticle and oomoldic pores) decreases in this unit. Intraparticle, moldic and intercrystalline pores are the most frequent types of porosity. Intercrystalline pores are most common in the upper parts and intraparticle and moldic pores in the lower parts. Minor dissolution caused increase in the effective connection of primary pores due to lack of anhydrite cement. The presence of intercrystalline porosity as a result of dolomitisation enhanced reservoir quality by connecting the primary pores. Planar-e dolomites are most common in this unit.

As in the K4 unit, fracturing has had a minor influence on porosity generation. Permeability has been increased in rare cases when factures are open. Stylolites commonly created by chemical compaction, are not common in this unit.

K1 Unit: This unit is about 82.5 m thick and in general consists of dolomites and limy dolostones with some limestones at the middle. Average porosity is 5.78% and average permeability is 15.22 md. Anhydrite is common in this unit mainly as of blocky cements. It is a very important diagenetic event as generally caused complete pore occlusion (plugging) and drastically diminishing poroperm values. Anhydrite cementation and compaction features such as stylolite and solution seams, are the most frequent diagenetic processes observed in K1. Filled fractures were observed in some parts of the unit and fracture-filling anhydrite reduced overall poroperm values. Compaction-related effects reduced interparticle and intraparticle porosity in K1 facies. Moldic and interraparticle are the most frequent types of porosity in limestone facies, while intercrystalline porosity is mainly common in the dolomitic facies.

Ternary Plot Construction

Classification of carbonate pore spaces developed by Choquette and Pray (Choquette and Pray, 1970) is applicable yet, but in recent years this classification has come under criticism (Lucia, 1999 and Lucia, 2007). This is due to the fact that this method emphasises on the importance of pore space genesis and thus its proposed pore classes are genetic and not petrophysical (Lucia, 1999 and Machel, 2005). However, consideration and modification of this classification in a new manner could provide insights into both genetical and petrophysical aspects of porosity in the carbonate reservoirs.

Because carbonate diagenesis has significant effects on the porosity creation, destruction or modification (Ahr, 2008, Machel, 2005 and Moore, 2001), as well as pore types, thus identification of the pore types and their changing trends will lead to recognition of major diagenetic processes and trends. Such a method has been successfully applied to the carbonates previously (Kopaska-Merkel and Mann, 1994 for example). Ternary pore plots provide quantitative information on the shapes and origins of pore-system elements. They provide insight into a variety of geological problems, including (I) identification of flow, baffle and barrier units; i.e., stratigraphic intervals that have significantly different fluid-flow properties; (2) recognition of diagenetic processes and trends; and (3) identification and quantification of pore facies or characteristic kinds of pore systems that may be of regional extent (Kopaska-Merkel and Mann, 1994).

Here, by considering all previous studies, a new ternary plot has been introduced. In fact, by using modified genetic classification for carbonate rock porosity (Choquette and Pray, 1970), a ternary diagram whose apexes are carbonate pore types (ternary pore plots) is employed to summarize quantitative data derived from petrographical digital point counting of thin sections (Jmicrovision v1.2 Image Analysis Software) (Fig. 6). Fortunately, diagenetic changes are not difficult to identify in thin sections. Thin-section point-count data are inexpensive and easy to collect, and could be used to guide more expensive engineering analyses.

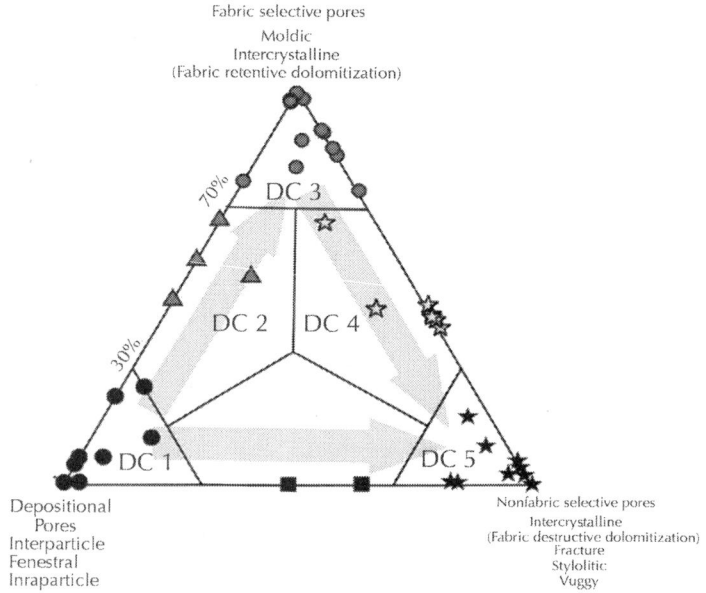

Figure 6: Ternary plot for pore types classification. Diagenetic trends based on porosity variations are depicted. Various diagenetic processes and their impact on this trend are apparent. Depositional pores will be modified with dissolution, compaction, dolomitisation, cementation, fracturing and stylolitisation. Each of these processes make specific trend of variation in porosity type and value. Studied samples were shown with different symbols. DC is "Diagenetic Class.

Fig. 6 shows this classification and diagenetic processes that created them. Each domain in this plot is clarified by a number that shows different degree of porosity evolution from depositional (Class 1) to completely non-fabric selective (absolutely diagenetic in origin) (Class 5). This approach simplifies the interpretation of pore systems by focusing only on three major components and by displaying these data on a simple graphic plot that makes trends and clustering of samples clear. The utility of this technique for carbonate reservoir-rock studies has been demonstrated with examples from various parts of the world (Kopaska-Merkel and Mann, 1994 for example).

In this study, three apexes of the ternary pore plots are attributed to: (1) depositional pores (interparticle, fenestral, intraparticle); (2) fabric-selective pores (moldic and intercrystalline as well as fabric-retentive dolomitisation); and (3) non-fabric selective pores ({intercrystalline and fabric-destructive dolomitisation}, fracture, stylolite, vug) (Fig. 6). Rocks that have no primary or secondary diagenetic-mediated porosity (no visible porosity) have not any reservoir interest. These three porosity types are the most common types of pores in the Kangan and Dalan formations in South Pars gas field and account for more than 95% of total porosity in the studied thin sections. In this scheme of pore systems classification which applied to the studied carbonate reservoirs, there are five classes (Fig. 6). According to this classification scheme for the porosity, primary pores are mainly depositional (up to 70%). It means that the effect of diagenesis on the porosity formation is subordinate and intrinsic porosities are present (Class 1). Studied intervals exhibit four basic depositional pore types: (1) interparticle, (2) intraparticle, (3) shelter or keystone, and (4) fenestral pores. Shelter porosity is not common in studied Kangan and Upper Dalan intervals. So, the diagenetic Class 1 contains interparticle, intraparticle and fenestral porosities. Then diagenesis has altered primary porosities in two different ways. In the fabric-selective pore alteration scheme, porosity destruction could be due to cementation or compaction. Dissolution also could be take place after deposition. By development of fabric-selective diagenesis in this way, Class 2 is created which shows between 30 to 70% of the original depositional pores.

Progressive evolution of pores by fabric-selective diagenesis creates new porosities in the rock body which is mainly diagenetic but fabric-selective (Class 3). Class 3 typically forms in particle-supported carbonate rocks that have been partially to wholly cemented in the shallow-marine-phreatic diagenetic environment and whose particles have undergone partial to complete dissolution. The moldic pore size is determined by former particle sizes and relatively high pore/throat size (aspect) ratios. Pore size and shape are determined by the former particle boundaries; pores are commonly spherical to elliptical and several hundred micrometers across because the most common particles were peloids and ooids. Permeability values in these samples increase only slightly with increasing porosity values.

By further diagenetic alterations, the new pores mainly cross-cut primary fabrics, and thus the non-fabric selective pores (Class 4) are created. In the extreme diagenetic alterations all new pores in the rock body are completely non-fabric selective (Class 5). In the latter case, porosities could be produced due to non-fabric selective dissolution and dolomitisation, fracturing or stylolitisation. Primary rock fabric has little influence on the distribution of intercrystalline porosity. Pore volume in the Class 5 is typically less than in the Class 3. However the aspect ratio is also smaller and a certain porosity value in this class typically corresponds to a higher permeability than in the moldic dominated strata. Although the range of porosity and mean porosity of the Class 5 is less than that of the other classes, the mean and maximum permeability is higher in the fabric-destructive intercrystalline dominated pores of this class. Fracuring could be an important factor for permeability increase in this zone. It is important to note that depositionally-mediated pore types should be distinguished from diagenetically-mediated, based on the detailed thin section studies (Fig. 6). Variations of diagenetic classes (DC) versus lithology were shown in each unit (Fig. 7).

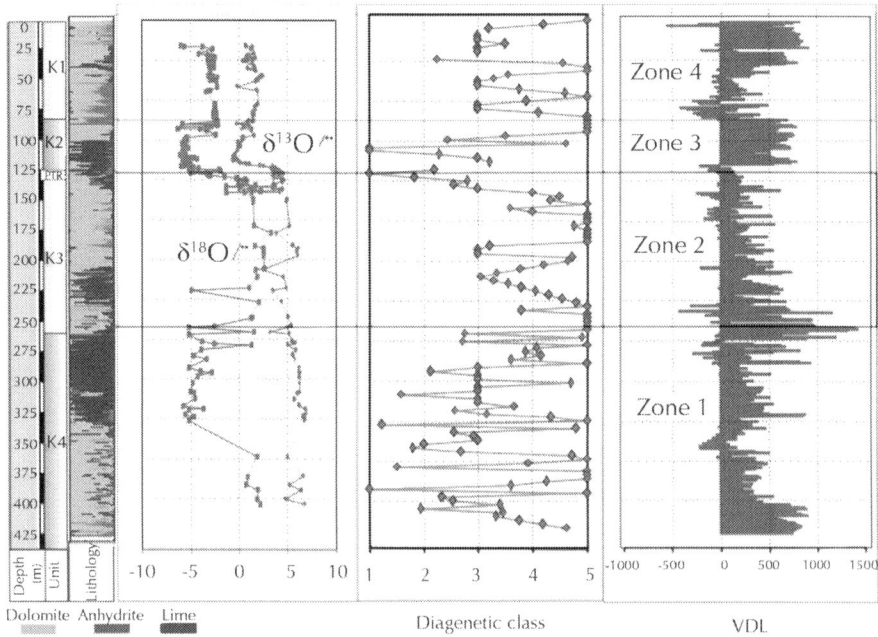

Figure 7: Lithological percentage column, carbon and oxygen stable isotopes, diagenetic classes and velocity-deviation log for well OFA_1 illustrated beside each other. PTB boundary is seen on the units column. Four zones have been highlighted.

Reservoir Quality

Here, the reservoir quality prediction is addressed by identification of the main porosity and permeability controlling processes (depositional and diagenetic) and evaluating how these processes vary on the scale of interest (well, reservoir, basin). The combination of porosity and permeability data in terms of RQI (reservoir quality index), is convenient for use with the routine core analysis data that gives a tremendous advantage in addressing the reservoir quality in various scales. The concept of Amaefule et al.'s (Amaefule et al., 1993) method is based on the calculation of this term, defined as follows:

$$RQI = 0.0314\sqrt{\frac{k}{\Phi}}$$

(1)

where: RQI: Reservoir quality index, μm; k: Permeability, md; Φ: Porosity, volume fraction.

Reservoir quality index of studied formations has been shown in Fig. 8. This value is a good representative of two basic reservoir quality indices (porosity and permeability) but size distribution of throats is not clear based on this index. Such a distribution could reveal the impact of diagenetic events on these two important factors.

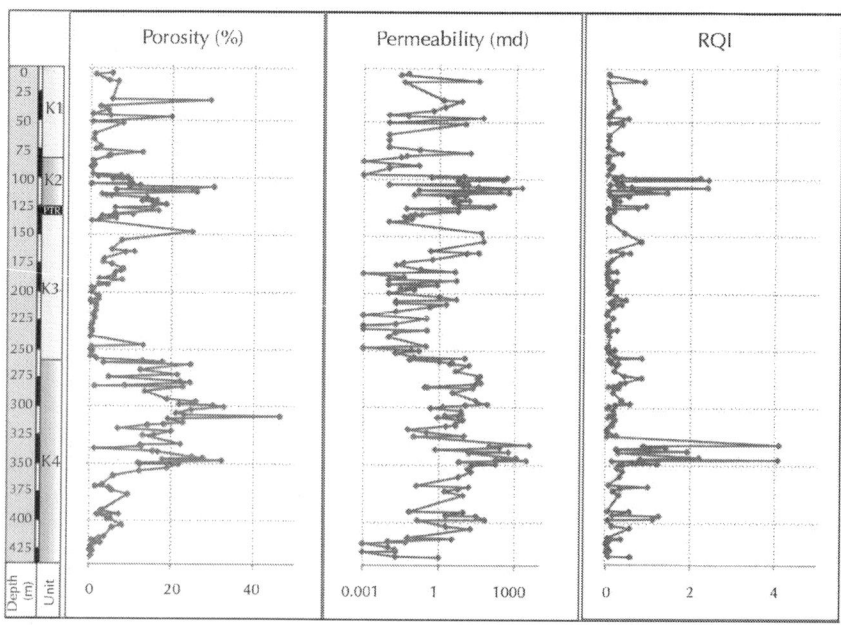

Figure 8: Porosity, permeability and RQI of Permo-Triassic Kangan and Dalan formations in well OFA_1 of South Pars gas field derived from 430 m of cores.

Pore-throat size distributions from mercury injection capillary pressure (MICP) tests are another factor used here for evaluation of reservoir quality and impact of diagenesis. The capillary pressure curve (CPC) shows that the distribution of pore-throat sizes (PTS) are in the range 0.01–92 µm. Fig. S3 shows MICP data from eight samples through K1-K4. As it could be seen in this figure, in K1 sample (Fig. S3a) dominant pore-throat size ranges from 0.6–0.8 µm with low porosity and permeability. In upper K2 sample (Fig. S3b) dolomitisation caused larger PTS with lower standard deviation. Lower K2 sample (Fig. S3c) has higher PTS dominantly range from 0.9-60 µm with higher porosity and permeability. The story of K1 unit repeats again for K3 sample (Fig. S3d) with lower mean PTS. In K3-K4 boundary with increasing rate of dissolution, dominant PTS varies toward higher values (Fig. S3e). In upper K4 unit (Fig. S3f, g) there is large PTS and high porosity and permeability. Lower K4 (Fig. S3h) has almost high permeability but porosity decreases effectively.

Oxygen and Carbon Isotopes

The carbon isotope data for three studied wells are illustrated in Fig. 9. Main statistical analysis of isotopic data in three wells has been shown in Table S1. ^{13}C values show relatively constant to decreasing trends in all three wells, from base of Permian to the PTB. Close to the PTB a sudden decrease occurs from 4.14‰, 5.12‰ and 3.21‰ to −0.95‰, −0.84‰ and −0.82‰ in wells OFA_1, OFA_2 and OFA_3, respectively. Upward in the section, after this shift, a subtle increase is clear in all wells.

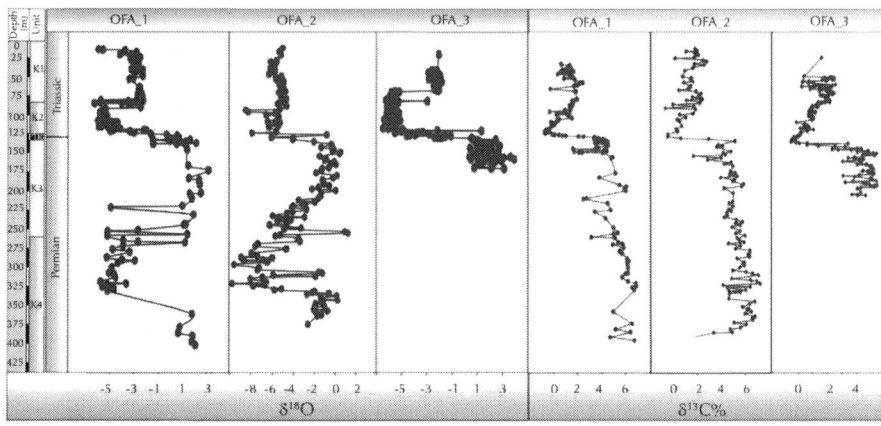

Figure 9: Carbon and oxygen stable isotope data in three studied wells.

Wells OFA_1 and OFA_2 have a nearly complete $\delta^{18}O$ data from all over the upper Permian to lower Triassic of Kangan and upper Dalan strata. Well OFA_3 has $\delta^{18}O$ data just near the boundary (Fig. 9). These three wells show a similar trend in $\delta^{18}O$ excursions. There are higher-frequency fluctuations in K3-K4 boundary and then a subtle increase occurs from lower K3 to the top of this interval (boundary between K2 and K3). This increase is clearer in OFA_2 well. A sudden decrease occurs at Permo-Triassic boundary (PTB). In general, $\delta^{18}O$ is lower in OFA_2 well. After the PTB, $\delta^{18}O$ varies from −7.28‰ to −4.91‰ until the K1-K2 boundary in this well.

Velocity-deviation Log

Combination of porosity, permeability, PTS and stable isotope data reveal quantitative reservoir properties, but pore geometry is another factor that controls reservoir quality in South Pars field. Wire-line logs are another source of information that could be used for understanding nature of pores in carbonates. The velocity-deviation log (VDL), which is calculated by combining the sonic log with the neutron-porosity or density log, provides a tool to obtain downhole information on the predominant pore type in carbonates. The log can be used to trace the downhole distribution of diagenetic processes and to estimate permeability trends (Anselmetti and Eberli, 1999) The VDL is calculated by first converting porosity-log data to a synthetic velocity log using a time-average equation (Wyllie and Gardner, 1956):

$$\frac{1}{V_{rock}} = \frac{1-\phi}{V_{matrix}} + \frac{\phi}{V_{fluid}}$$

(2)

The difference between the real sonic log and the synthetic sonic log can then be plotted as a velocity-deviation log. Because deviations are the result of the variability of velocity at certain porosity, the deviation log reflects the different rock-physical signatures of the different pore types.

Positive deviations indicate relatively high velocities in regard to porosity, and are caused mainly by porosity that is integrated in a framelike fabric of the rock, such as in intrafossil or moldic porosity. In the moldic pores, porosity deviations indicate intense diagenetic alterations, such as dissolution or precipitation, yielding moldic porosity and favoring re-precipitation of the dissolved material as pore-filling cement.

Zones with small deviations (±500 m/s or less) represent sections that follow the predictions by the time-average equation. These zones are dominated by either interparticle, intercrystalline, or high microporosity.

There are three possible explanations for zones with the negative deviation. First this negative deviation could be the result of caving or irregularities of the borehole wall. Another reason for negative deviation is fracturing. Despite the fact that fracture porosity has always been included in the secondary porosity (equivalent to high velocity or positive deviations) (Schlumberger, 1974), several studies showed that fracturing decreases velocities on both small (Anselmetti and Eberli, 1993 and Gardner et al., 1974) and large (Guadagno and Nunziata, 1993) scales. The large scale fractures can be detected with the logging tools and yield lower velocities than the undisturbed rock. In addition, buried fractures are generally closed or, in regard to total porosity, are relatively insignificant, so that the neutron porosity is not significantly reduced. As a result, fracturing produces negative deviations. These negative deviations could be also caused by a high content of free gas. Free gas would have a strong negative effect on the deviation log because gas drastically reduces Vp (Nur and Simmons, 1969) and results in a reduced neutron porosity reading due to the lower content of hydrogen in the fluid phase (Hilchie, 1982).

To produce a velocity-deviation log in the studied interval in well OFA_1, combination of neutron-porosity and porosity from density logs was used. Logs have been corrected for bad hole intervals and gas effect and then difference between real sonic and synthetic velocity was depicted.

DISCUSSION

In heterogeneous carbonate reservoirs, a successful study requires integration of various data sets and methods of data analysis into a unified, interdisciplinary approach. For such an analysis, porosity and permeability data are the most critical factors. Integration of porosity and permeability data in terms of RQI provide a good relationship between these properties. Fig. 10 shows a plot of permeability versus porosity data obtained from wells OFA_2 and OFA_3. As could be seen in this figure, K1 and K3 units have no high porosity or permeability and so no reservoir quality. There are no obvious, systematic differences in porosity-permeability distribution between these units. Distinction between fabric-selective pores and non-fabric selective ones with regards to porosity-permeability characteristics can be evaluated in Fig. 10, K2 and K4 units. Upper K2 samples have higher permeability with the same porosity as lower K2 ones. Dolomitised lower K4 samples show generally higher permeability for given porosity, believed to reflect the effect of fabric destructive dolomitisation. These observations show that dolomitization enhanced permeability in both fabric-selective and fabric-destructive form. Fig. 8 shows porosity, permeability and RQI of the well OFA_1. As it is expected, porosity and permeability are higher in K4 and K2 units. RQI following porosity and permeability data has higher values in K4 and K2.

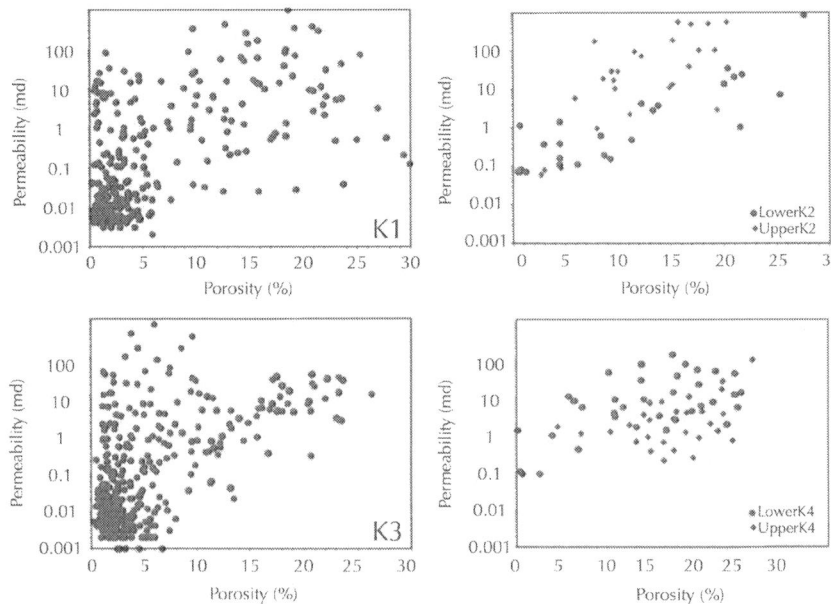

Figure 10: Plot of permeability versus porosity data obtained from wells OFA_2 and OFA_3.

Constructing a plot of RQI versus rock texture has been shown in Fig. 11. As shown, despite wide diversity, textural categories do not occupy distinctly separate RQI fields. This shows the effect of diagenesis on the reservoir characteristics in the field. This is in agreement with the previous works on South Pars field (Esrafili-Dizaji and Rahimpour-Bonab, 2009 for example). A plot of RQI versus DCs is constructed for better perception of relationships between RQI and diagenetic classes (Fig. 12). As it could be seen in this figure, DC4 has the highest reservoir quality with the average of 0.37. This is due to the fact that porosities in this zone are the result of dissolution that was enhanced by the subsequent dolomitisation. DC1 has the lowest reservoir quality with the average of 0.12 that reflects non-fabric selective pores (primary pores without important diagenetic imprint). In spite of this fact that Class 5 has relatively higher permeability than Classes 2 and 3, reservoir quality is not very high in this zone. This is due to low porosity values of this class.

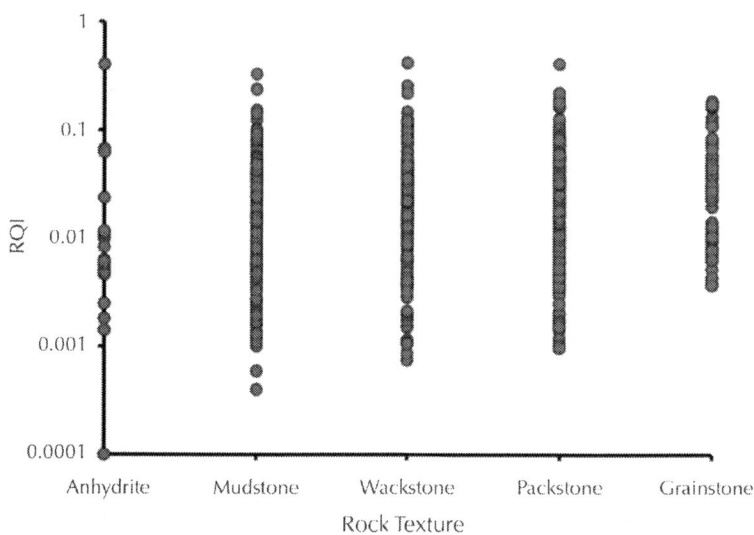

Figure 11: Plot of rock texture versus RQI shows no obvious relationships between two parameters.

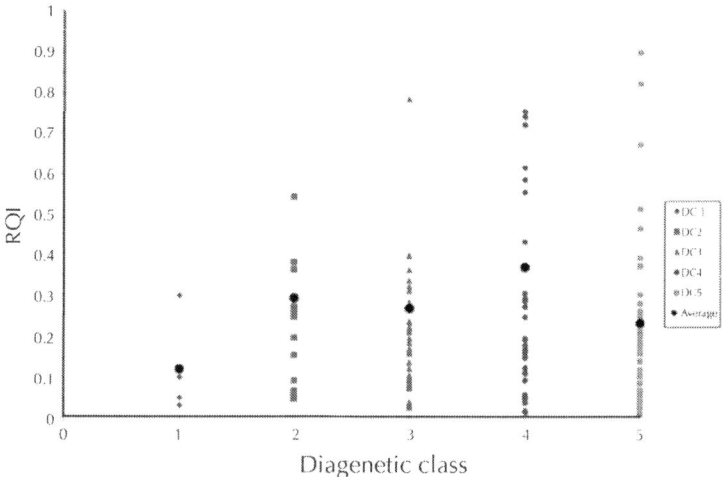

Figure 12: Plot of diagenetic classes versus RQI show that intervals with connected fabric-selective dissolution have the best reservoir quality and non-connected primary pores have not any interest in reservoir quality point of view. Refer to text for more explanation.

Pore-throat size distribution has been widely used for understanding the pore geometry and resulted reservoir quality (Bliefnick and Kaldi, 1996, Cerepi et al., 2003, Mancini et al., 2004 and Martin et al., 1997; Taghavi et al, 2004). In upper K2, non-fabric selective classes show moderate PTS with almost high permeability and moderate porosity. In lower parts of this unit lower diagenetic classes show creation of larger PTS caused by dissolution. The story repeats again in K4, but vice versa in upper and lower parts in comparison with K2. Decreasing PTS and porosity with minor change in permeability shows the effect of dolomitisation in lower K4 unit. It is in good agreement with DCs in reservoir zones.

Intervals with poor reservoir quality are characterized by platykurtic throat size distribution and there is obvious mode in CPC, which is shown in Fig. S3a and d. Upper K2 reservoir unit which is characterized by fabric destructive dolomitization, exhibits leptokurtic throat size distribution in Fig. S3b. Such distribution shows that pore distribution is nearly uniform and pore throats have almost the same sizes, more than 1 μm. The same trend could be seen in lower K2 unit but complementary studies (petrography and VDL log, for example) show that dissolution is the ominant connecting factor in this unit. In upper K4, where dissolution is the dominant factor for porosity and permeability creation, pore-throat size exhibit platykurtic, polymodal distribution (Fig. S3f). With increasing dolomite volume in this part, the curve tends to be flattened (Fig. S3g). Dolomites create different size connections from dissolution and so curve will be flattened. The CPC of lower K4 unit is characterizing with mesokurtic, biomodal PTS distribution. Two modes are very close to each other. Such distribution is due to fabric retentive to selective dolomitization of lower K4 unit. Same size dolomite crystals of this unit create PTS distribution shown in Fig. S3h.

Distribution of pore-throat sizes shows that pore system of diagenetic classes from 1 to 5 differs fundamentally from one another. This difference has a major effect on fluid-flow properties of pore systems dominated by one or the other of these systems. Moldic and fabric-selective dolomitization pore systems are heterogeneous on a microscopic scale because they consist of large pores that are connected to one another by distinctly different (commonly finer) pore systems. Note that pore sizes, but not necessarily pore-throat sizes, are inherently heterogeneous in the Class 3. By contrast, Class 5 pore system is essentially homogeneous, because they are developed in rock fabrics that tend to consist of unimodal leptokurtic distributions of dolomite crystals that are also about the same shape. It should be mentioned that there is no considerable fracturing in studied intervals in South Pars field and so permeability has no anomalously variations.

Stable isotope analysis is one of the best methods for diagenesis studies in carbonate strata (Al-Aasm and Azmy, 1996, Ehrenberg et al., 2006). Observations of stable isotope excursion trough Kangan and upper Dalan cored interval in South Pars field are indicative of ^{13}C depletion over the PTB, concomitant with the worldwide ^{13}C excursion at this time. The magnitude of this shift is also analogous to the other typical PTB sections in the world (e.g., Dolenec et al., 2001, Heydari and Hassanzadeh, 2003, Lehrmann et al., 2003,Korte et al., 2004 and Rahimpour-Bonab et al., 2009). These findings suggest that a large input of ^{12}C-enriched carbon into the ocean-atmosphere system has occurred and may have caused a global environmental change, probably related to this greatest mass extinction in the Phanerozoic. The increasing trend toward the top (relative to PTB) has been related to restoration of primary productivity (Rahimpour-Bonab et al., 2009), while decreasing trend from bottom of the section toward the PTB could be ascribed to reoxidation of previously stored ^{12}C-enriched organic material caused by eustatic regression (Erwin, 1994, Heydari et al., 2000, Horacek et al., 2007 and Loydell, 2007).

Heydari et al. (2000) related PTB isotope changes to the eustatic regression at the end Permian time. They have mentioned negative isotopic [18]O values and recrystallization of lime mud matrix. In contrast, Korte et al. (2004) believe that the Late Permian sea-level fall did occur on the Russian Platform, in North America and Siberia, but a time-equivalent large-scale transgression happened in China, Iran, western Tethys and in the Zechstein Basin of central

Europe. Evidences represented by thin sections and SEM studies as well as wire-line logs and stable isotopes, place more emphasis on the diagenetic origin for these isotopic anomalies. Domination of greenhouse conditions over the Early Triassic (Kidder and Worsley, 2004 and Retallack, 1999) cause low (light) $\delta^{18}O$ through K2 unit.

In spite of some little differences, all three wells show similar isotope trends with well OFA_1. Comparison of stable isotopes (especially oxygen) of this well with diagenetic classes determined by detailed thin section examinations (Fig. 7) reveals some similarity. As shown in this figure, there are four distinguishable zones. Through whole of K4 and the base of the K3, diagenetic classes disperse to the fabric-selective and non-fabric selective classes (2, 3, 4 and some 5 classes). These classes are mainly fabric-selective pores which their poroperm values increased by the later fabric-destructive dolomitisation and/or dissolution. Petrographic and SEM evidences show that most of K4 unit has intercrystalline, moldic and some vuggy porosity. Regarding the nature of pore types in this interval, it could be concluded that the primary pores were mainly enhanced by the fabric-selective dolomitisation and dissolution, but permeability improved by the non-fabric selective dolomitisation and dissolution. Decreasing trend in the $\delta^{18}O$ values accompanied with the limy lithology in the upper parts of K4 indicates domination of dissolution while the large scale dolomitisation is predominating in the lower part.

In K3 unit, with increasing $\delta^{18}O$ and $\delta^{13}C$ values, diagenetic classes shift to the higher degrees (Fig. 7, zone 2), i.e., the non-fabric selective diagenetic pores dominate. As is mentioned earlier, development of meteoric and micrite cement in K3 unit has occluded pore-throats and so reservoir quality is not high. In comparison with the planar-e dolomites of K2 and K4, poroperm values in K3 planar-s dolomites are not high. There is a high permeable zone just below the PTB that is due to grain-dominated facies. However, because of diagenetic effects these values show a dramatic decline passing the PTB interval.

After this important interval, toward K2 unit, zone 3 shows more negative $\delta^{18}O$ values. With depletion in the $\delta^{18}O$ values, diagenetic classes decrease toward the lower ones including 1, 2 and 3 (especially 1 and 3). Considering lime and dolomite lithology of K2 interval and nature of the pore types, primary porosity was enhanced by the fabric-selective dissolution and recrystallization resulted in concurrent permeability increase and reservoir quality improvement. In addition, parallel decrease in $\delta^{18}O$ values (Fig. 7), as a result of meteoric diagenesis, supports this effect.

Zone 4 is from top of the K2 to the uppermost interval (about 100 m from top of K2). This zone includes three diagenetic classes of 3, 4 and 5. There are no sharp variations in ^{18}O and ^{13}C values in this interval. According to diagenetic classes scheme (Fig. 6), in this zone porosity was developed mainly as intercrystalline, fracture and stylolite. Thus, observed pores are not primary and fabric-selective. Porosity volume is not high in this zone. In addition to this fact, petrographical and SEM observations showed that there is considerable volume of anhydrite in this interval that occluded most pores. This resulted in the reservoir quality loss.

Understanding the relationship between pore type and porosity and permeability in a general view is a major challenge in the evaluation of carbonate reservoirs. Here, VDL in combination with foregoing parameters employed for this purpose. In VDL point of view, the lowermost studied interval (K4) is a high quality zone with 0.85 N/G (Rahimpour-Bonab et al., 2009). VDL has average of 320 m/s in this interval that shows the dominance of the connected porosities. Ranging diagenetic classes between 2 and 5 shows that both fabric and non-fabric selective dissolution and dolomitisation are the important causes for the improvement of reservoir quality. Regarding nature of the pore types of this interval and VDL log variations, it could be concluded that primary pores were mainly enhanced by the fabric-selective dolomitisation and dissolution, while permeability improved by the non-fabric selective dolomitisation and dissolution.

In spite of well-connected porosities in the K3 unit deduced from VDL and diagenetic classes, high rate of cementation occluded pore spaces and filled pore throats imparting low reservoir quality. Generally, the VDL remains positive in the K2 while the diagenetic classes are in the range of 1 to 3. This indicates the presence of primary (Class 1) and fabric-selective diagenetic (Classes 2 and 3) pores. Domination of diagenetic Class of 5 in the uppermost of K2 is confirming the role of dolomitisation in this interval. In contrast to the K4 unit, dolomitisation is more effective in the upper parts while dissolution created reservoir quality in the lower parts of K2. Although dolomitisation decreases downward, pores are still connected that is reflected in the RQI values (Fig. 8). Fabric-selective dissolution in this interval produced important moldic porosity that has moderate connectivity. There is an increasing trend in the velocity deviation from base of the K1 unit to the top of this interval (Fig. 8). Positive values in this interval support the petrographic observations that reveal pore-filling anhydrites.

Ehrenberg et al. (2006) found similar porosity and cementation in Finnmark Formation. They showed that primary intergranular and intrafossil pores have been filled by calcite cement. Dolostones contain two dominant pore types: intercrystalline pores between dolomite crystals and moldic pores representing former bioclasts. Their study revealed that reservoir compartmentalization by the formation of tight limestone barriers is largely a burial diagenetic process involving calcite cementation locally produced by chemical compaction as could be seen here in K1 and K3 units. Eichenseer et al. (1999) believe that in Pinda Group intercrystalline porosity allows connection between moldic pores and enhances reservoir quality. Intracrystalline porosity is frequent, although volumetrically of minor importance. This is the case in lower K4 unit in our study. The same diagenetic trend has been seen by Pomar and Ward (1999) in Miocene deposits of Mallorca Island in Spain. Woody et al. (1996) note that in Cambrian–Ordovician dolomites from core and outcrop throughout southeastern Missouri, permeability displays a strong logarithmic increase with increasing porosity in planar-e dolomite. Permeability in planar-s dolomite is lower than in planar-e dolomite and does not increase as rapidly with increasing porosity. High-pressure mercury capillary pressure curves and SEM data show that planar-e dolomites are comprised of uniform well-interconnected pore systems. This is the same for K3 and K4 dolomites. Dolomites of K4 are planar-e where as K3 dolomites are planar-s in shape. Similar diagenetic trend and phases with porosity development were reported by Ehrenberg (2006), Esrafili-Dizaji and Rahimpour-Bonab (2009), Heasley et al. (2000), Kopaska-Merkel (1992), and Machel (2005).

CONCLUSIONS

The Permo-Triassic carbonates in the South Pars gas field have complicated diagenetic history including marine, meteoric and burial diagenesis which makes it difficult to evaluate their reservoir properties. Reservoir quality in this field is controlled by the spatial heterogeneity in diagenetic cementation, dissolution, dolomitisation and resultant porosity development. Here, an integrated approach using porosity-permeability data in terms of RQI, mercury intrusion data, stable isotopes, petrographic studies, SEM and VDL log was used for investigating the effect of diagenesis on the reservoir quality of Kangan and Dalan formations.

There are two main reservoir zones in the Permo-Triassic interval of the studied reservoir including K2 and K4. The reservoir quality of these two zones has been mentioned previously in many studies but the clear relationship between these reservoirs and their diagenetic events in various parts had not been understood yet.

K4 is the best reservoir zone in this field with 0.85 N/G. The maximum values of porosity (average of 15.61%) and RQI are encountered in this zone. Diagenetic classes ranging mainly from 3 to 4 in the upper parts of the unit beside the higher VDL values, limy lithology and depleted ^{18}O values all confirm the role of fabric selective dissolution in the upper parts. In the lower parts of K4, pervasive dolomitisation and intercrystalline porosity that is reflected on the diagenetic classes and VDL log, along with the petrographic observations are the evidences for the creation of secondary connectivity of the pores. In addition, primary porosities were connected following diagenetic alterations. Most of the pores in K3 are secondary in origin but pervasive cementation caused low reservoir quality that is reflected on RQI values. The nature of pores (mainly stylolite and fractures) in this interval is another cause of low reservoir quality.

K2 has high porosity, permeability and RQI values. With average of 38.81 md, this unit has the best permeability and is one of the two main reservoir zones in the South Pars field. Depletion of stable isotope values, diagenetic classes and VDL confirm that the good quality of this zone is due to primary pores enhanced by dolomitisation in the upper parts and dissolution in the lower parts. Comparison of diagenetic classes in upper part of K2 and lower part of K4 shows some little differences. While both parts are dolomitised, in upper K2 unit non-fabric selective pores are dominant and fabric destructive dolomitisation is the main cause of high reservoir quality. In comparison, lower K4 has more fabric-selective pores that have been connected by fabric retentive to selective dolomitisation. Diagenetic cements are heterogeneously distributed throughout the reservoir and have a maximum effect on the dominant pore network in K3 and K1 unit. Pore occluding diagenetic anhydrite is consistently most abundant in these units. Diagenetic studies in terms of diagenetic classes show secondary pores in the K1 unit. In spite of this fact, moldic nature of some pores and pervasive anhydrite cementation filled pores and occluded pore-throats. As it mentioned, RQI has low values in this zone and good reservoir intervals are not present. Results of this study show that good conclusions will achieved through integration of various data and geological concepts.

ACKNOWLEDGMENTS

The vice-president of Research and Technology of the University of Tehran provided financial support for this research, which we are grateful. We also extend our appreciation to the POGC (Pars Oil and Gas Company of Iran) for sponsoring, data preparation, and permission to publish this paper. We would like to thank Mrs. Naderi for her detailed and helpful reviews and comments.

REFERENCES

1.   Aali, J., Rahimpour-Bonab, H., Kamali, M.R., 2006. Geochemistry and origin of the world's largest gas field from Persian Gulf, Iran. J. Petrol Sci. Engin. 50, 161–175.

2.   Abid, I., Hesse, R., 2007. Illitising fluids as precursors of hydrocarbon migration along transfer and boundary faults of the Jeanne d'Arc Basin offshore Newfoundland, Canada. Mar. Pet. Geol. 24, 237–245.

3.   Ahr, W.M., 2008. Geology of carbonate reservoirs, the identification, description and characterisation of hydrocarbon reservoirs in carbonate rocks. Wiley Publication, New Jersey, 278 p.

4.   Al-Aasm, I.S., Azmy, K.K., 1996. Diagenesis and evolution of microporosity of Middle–Upper Devonian Kee Scarp reefs, Norman wells, Northwest territories, Canada: petrographic and chemical evidence. AAPG Bull. 80, 82–100.

5.   Al-Aswad, A.A., 1997. Stratigraphy, sedimentary environment and depositional evolution of the Khuff Formation in south-central Saudi Arabia. J. Pet. Geol. 20, 1–20.

6.   Al-Husseini, M.I., 2000. Origin of the Arabian Plate Structures: Amar Collision and Najd Rift. Geo. Arabia 5, 527–542.

7.   Alsharhan, A.S., 2006. Sedimentological character and hydrocarbon parameters of the Middle Permian to Early Triassic Khuff Formation, United Arab Emirates. GeoArabia 11, 121–158.

8.   Alsharhan, A.S., Kendall, C., 2003. Holocene coastal carbonates and evaporites of the southern Arabian Gulf and their ancient analogues. Earth-Science Rev. 61, 191–243.

9.   Alvarez, N.O.C., Roser, B.P., 2007. Geochemistry of black shales from the Lower Cretaceous Paja Formation, Eastern Cordillera, Colombia: Source weathering, provenance, and tectonic setting. J. South Am. Earth Sci. 23, 271–289.

10. Amaefule, J.O., Altunbay, M., Tiab, D., Kersey, D.G., Keelan, D.K., 1993. Enhanced Reservoir Description: Using core and log data to identify Hydraulic (Flow) Units and predict permeability in uncored intervals/wells. SPE 26436. Presented at 68th Ann. Tech. Conf. and Exhibition, Houston, TX.

11. Anselmetti, F.S., Eberli, G.P., 1993. Controls on sonic velocity in carbonates. Pure Appl. Geophys. 141, 287–323.

12. Anselmetti, F.S., Eberli, G.P., 1999. The velocity-deviation log: a tool to predict pore type and permeability trends in carbonate drill holes from sonic and porosity or density logs. AAPG Bull. 83, 450–466.

13. Baron, M., Parnell, J., Mark, D., Carr, A., Przyjalgowski, M., Feely, M., 2008. Evolution of hydrocarbon migration style in a fractured reservoir deduced from fluid inclusion data, Clair Field, West of Shetland, UK. Mar. Pet. Geol. 25, 153–172.

14. Bliefnick, D.M., Kaldi, J., 1996. Pore geometry: control on reservoir properties, Walker Creek Field, Columbia and Lafayette Counties, Arkansas. AAPG Bull. 80, 1027–1044.

15. Bordnave, M.L., 2008. The origin of the Permo-Triassic gas accumulations in the Iranian Zagros foldbelt and contiguous offshore area: a review of the Paleozoic petroleum system. J. Pet. Geol. 31, 3–42.

16. Cerepi, A., Barde, J.P., Labat, N., 2003. High-resolution characterisation and integrated study of a reservoir formation: the Danian carbonate platform in the Aquitaine Basin (France). Mar. Pet. Geol. 20, 1161– 1183.

17. Choquette, P.W., Pray, L.C., 1970. Geologic nomenclature and classifi- cation of porosity in sedimentary carbonates. AAPG Bull. 54, 207– 250.

18. Dolenec, T., Lojen, S., Ramovs, A., 2001. The Permian-Triassic boundary in western Slovenia (Idrijca Valley section): magnetostratigraphy, stable isotopes, and elemental variations. Chem. Geol. 175, 175–190.

19. Ehrenberg, S.N., 2006. Porosity destruction in carbonate platforms. J. Pet. Geol. 29, 41–52.

20. Ehrenberg, S.N., Aqrawi, A.A.M., Nadeau, P.H., 2008. An overview of reservoir quality in producing Cretaceous strata of the Middle East. Pet. Geosci. 14, 307–318.

21. Eichenseer, H.T., Walgenwitz, F.R., Biondi, P.J., 1999. Stratigraphic control on facies and diagenesis of dolomitized oolitic siliciclastic ramp sequences (Pinda Group, Albian, offshore Angola). AAPG Bull. 83, 1729–1758.

22. Elias, A.R., Ros, D.L.F., Mizusaki, A.M., Anjos, S.M., 2004. Diagenetic patterns in eolian/coastal sabkha reservoirs of the Solimoes Basin, northern Brazil. Sed. Geol. 169, 191–217.

23. Erwin, D.H., 1994. The Permo-Triassic extinction. Nature 367, 231–236.

24. Esrafili-Dizaji, B., Rahimpour-Bonab, H., 2009. Effects of depositional and diagenetic characteristics on carbonate reservoir quality: a case study from the South Pars gas field in the Persian Gulf. Pet. Geosci. 15, 1–22. Gardner, G.H.F., Gardner, L.W., Gregory, A.R., 1974. Formation velocity and density: the diagnostic basics for stratigraphic traps. Geophysics 39, 770–780.

25. Guadagno, F.M., Nunziata, C., 1993. Seismic velocities of fractured carbonate rocks (southern Apennines, Italy). Geophys. J. Int. 113, 739–746.

26. Halley, R.B., Harris, P.M., 1979. Freshwater cementation of a 1000-yearold-oolite. J. Sed. Pet. 49, 969–988.

27. Heasley, E.C., Worden, R.H., Hendry, J.P., 2000. Cement distribution in a carbonate reservoir: recognition of a palaeo oil-water contact and its relationship to reservoir quality in the Humbly Grove Field, onshore, UK. Mar. Pet. Geol. 17, 639–654.

28. Heydari, E., Hassanzadeh, J., 2003. Deev Jahi model of the PermianTriassic mass extinction: a case study for gas hydrates as the main cause of biological crisis on earth. J. Sed. Geol. 163, 147–163.

29.   Heydari, E., Hassanzadeh, J., Wade, W.J., 2000. Geochemistry of central tethyan Upper Permian and Lower Triassic strata, Abadeh region, Iran. J. Sed. Geol. 137, 85–89.

30.   Hilchie, D.W., 1982. Advanced well log interpretation. Golden, Colorado, 300 p.

31.   Horacek, M., Brandner, R., Abart, R., 2007. Carbon isotope record of the P/T boundary and the Lower Triassic in the Southern Alps: evidence for rapid changes in storage of organic carbon. Paleogeogr. Palaeoclimatol. Palaeoecol. 252, 347–354.

32.   Insalaco, E., Virgone, A., Courme, B., Gaillot, J., Kamali, M., Moallemi, A., Lotfpour, M., Monibi, S., 2006. Upper Dalan Member and Kangan Formation between the Zagros Mountains and offshore Fars, Iran: depositional system, biostratigraphy and stratigraphic architecture. GeoArabia 11, 75–176.

33.   James, N.P., Choquette, P.W., 1984. Diagenesis of Limestones; the meteoric diagenetic environment. Geosci. Can. 11, 161–194.

34.   Kashfi, M.S., 2000. Greater Persian Gulf Permian–Triassic stratigraphic nomenclature requires study. Oil and Gas Journal 6, 36–44.

35.   Kidder, D.L., Worsley, T.R., 2004. Causes and consequences of extreme Permo-Triassic warming to globally equable climate and relation to the Permo-Triassic extinction and recovery. Paleogeogr. Palaeoclimatol. Palaeoecol. 203, 207–237.

36.   Kirmaci, M.Z., 2008. Dolomitisation of the Late Cretaceous–Palaeocene platform carbonates, Go ̈lko ̈y (Ordu), eastern Pontides, NE Turkey. Sedimentary Geology 203, 289–306.

37.   Konert, G., Afif, A.M., AL-Hajari, S.A., Droste, H., 2001. Palaeozoic stratigraphy and hydrocarbon habitat of the Arabian Plate. GeoArabia 6, 407–442.

38.   Kopaska-Merkel, D.C., 1992. Geologic setting, petrophysical characteristics, and regional heterogeneity patterns of the

Smackover in Southwest Alabama. In: Geological Survey of Alabama

39. Kopaska-Merkel, D.C., Mann, S.D., 1994. Classification of lithified carbonates using ternary plots of pore facies-examples from the Jurassic Smackover Formation, Carbonate microfabrics symposium proceedings, Texas A and M University, College Station, Texas. In: Frontiers in Sedimentary Geology Series, p. 265–277.

40. Korte, C., Kozur, H.W., Joachimski, M.M., Strauss, H., Veizer, J., Schwark, L., 2004. Carbon, sulfur, oxygen and strontium isotope records, organochemistry and biostratigraphy across the Permian/Triassic boundary in Abadeh, Iran. Geol. Rundsch 93, 565–581.

41. Lehrmann, D.J., Payne, J.L., Flix, S.V., Dilllet, P.M., 2003. Permian-Triassic boundary sections form shallow marine carbonate platforms of the Nanpanjiang basin, south China: implications for oceanic conditions associated with the end Permian extinction and its aftermath. Palaios 18, 138–152.

42. Loydell, D.K., 2007. Early Silurian positive d13C excursions and their relationship to glaciations, sea-level changes and extinction events. Geol. J. 42, 531–546.

43. Lucia, F.J., 1999. Carbonate reservoir characterization. Springer, 226 p.

44. Lucia, F.J., 2007. Carbonate reservoir characterisation: an integrated approach, 2nd. edition. Springer, 336 p.

45. Machel, H.G., 2005. Investigations of burial diagenesis in carbonate hydrocarbon reservoir rocks. Geosci. Can. 32, 103–128.

46. Mancini, E.A., Llinas, J.C., Parcell, W.C., Aurell, M., Badenas, B., Leinfelder, R.R., Benson, D.J., 2004. Upper Jurassic thrombolite reservoir play, northeastern Gulf of Mexico. AAPG Bull. 88, 1573–1602.

47. Martin, A.J., Solomon, S.T., Hartmann, D.J., 1997. Characterization of petrophysical flow units in carbonate reservoirs. AAPG Bull. 81, 734–759.

48. Moore, C.H., 2001. Carbonate reservoirs: porosity evolution and diagenesis in a sequence stratigraphic framework. Amsterdam, Elsevier, 460 p.

49. Moradpour, M., Zamani, Z., Moallemi, S.A., 2008. Controls on reservoir quality in the Lower Triassic Kangan Formation, Southern Persian Gulf. J. Petrol. Geol. 31, 367–386.

50. Nur, A., Simmons, G., 1969. The effect of saturation on velocity in low porosity rocks. Earth Planet. Sci. Lett. 7, 183–193.

51. Pomar, L., Ward, W.C., 1999. Reservoir-scale heterogeneity in depositional packages and diagenetic patterns on a reef-rimmed platform, Upper Miocene, Mallorca, Spain. AAPG Bull. 83, 1759–1773.

52. Rahimpour-Bonab, H., 2007. A procedure for appraisal of a hydrocarbon reservoir continuity and quantification of its heterogeneity. J. Petrol. Sci. Engin. 58, 1–12.

53. Rahimpour-Bonab, H., Asadi-Eskandar, A., Sonei, A., 2009. Controls of Permian-Triassic Boundary over Reservoir Characteristics of South Pars Gas Field, Persian Gulf. Geol. J. 44, 341–364.

54. Rahimpour-Bonab, H., Esrafili-Dizaji, B., Tavakoli, V., 2010. Dolomitization and anhydrite precipitation in Permo-Triassic carbonates at the South Pars gas Field, Offshore Iran: controls on reservoir quality. J. Pet. Geol. 33, 43–66.

55. Retallack, G.J., 1999. Postapocalyptic greenhouse paleoclimate revealed by Earliest Triassic paleosols in the Sydney Basin, Australia. Geol. Soc. Am. Bull. 111, 52–70.

56. Schlumberger, 1974. Log interpretation-applications, Vol. 2. Schlumberger Limited, 116 p.

57. Stentoft, N., Lapinskas, P., Musteikis, P., 2003. Diagenesis of Silurian reefal carbonates, Kudirka oilfield, Lithuania. J. Petrol. Geol. 26, 381–402.

58. Szabo, F., Keradpir, A., 1978. Permian and Triassic stratigraphy Zagros Basin, Southwest Iran. J. Petrol. Geol. 1, 57–82.

59.   Taghavi, A.A., Mork, A., Emadi, M.A., 2006. Sequence stratigraphically controlled diagenesis governs reservoir quality in the carbonate Dehluran Field, Southwest Iran. Petrol. Geosci. 12, 115–126.

60.   Tucker, M.E., Bathurst, R.G.C., 1990. Carbonate Diagenesis. Blackwell Scientific Publications, Oxford, 320 p.

61.   Videtich, P.E., 1994. Dolomitization and H2S generation in the Permian Khuff Formation, Offshore Dubai, UAE. Carbonates and Evaporites 9, 42–57.

62.   Woody, R.E., Gregg, J.M., Koederitz, L.F., 1996. Dolomite: evidence from the Cambrian-Ordovician of southeastern Missouri. AAPG Bull. 80, 119–132.

63.   Wyllie, G., Gardner, G.H.F., 1956. Elastic wave velocities in heterogeneous and porous media. Geophysics 21, 41–70.

Chapter **5**

Integrated Multicomponent Solute Geothermometry

N. Spycher[a], L. Peiffer[a], E.L. Sonnenthal[a], G. Saldi[a],
M.H. Reed[b], and B.M. Kennedy[a]

[a]Earth Sciences Division, Lawrence Berkeley National Laboratory, Berkeley, CA 94720, United States

[b]Department of Geological Sciences, University of Oregon, Eugene, OR 97403, United States

ABSTRACT

The previously developed and well-demonstrated mineral saturation geothermometry method is revisited with the objective to ease its application, and to improve the prediction of geothermal reservoir temperatures using full and integrated chemical analyses of geothermal fluids. Reservoir temperatures are estimated by assessing numerically the clustering of mineral saturation indices computed as a function of temperature. The reconstruction of the deep geothermal fluid compositions, and geothermometry computations, are implemented into one stand-alone program, allowing unknown or poorly constrained input parameters to be estimated by numerical optimization using existing parameter estimation software. The geothermometry system is tested with geothermal waters from previous studies, and with fluids at various degrees of fluid–rock chemical equilibrium obtained from laboratory experiments and reactive transport simulations. Such an integrated geothermometry approach presents advantages over classical geothermometers for fluids that have not fully equilibrated with reservoir minerals and/or that have been subject to processes such as dilution and gas loss.

GRAPHICAL ABSTRACT

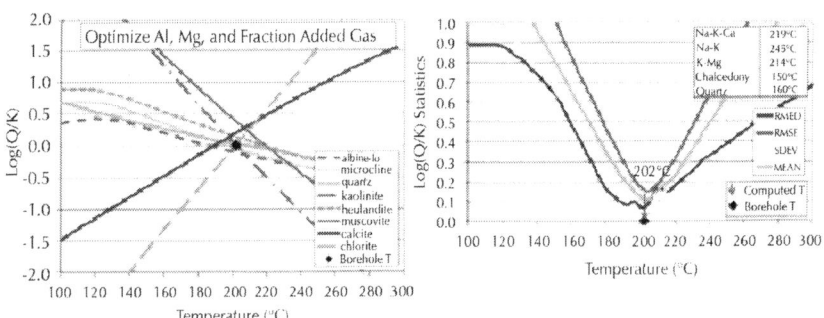

INTRODUCTION

Solute geothermometers have been used for decades to infer the temperature of deep geothermal reservoirs from analyses of fluid samples collected at ground surface from springs and exploration wells. The most commonly applied such geothermometers include those based on the concentration of silica (Fournier and Rowe, 1966, Fournier and Potter, 1982 and Fournier, 1977) and of sodium, potassium and calcium (Fournier and Truesdell, 1973, Fournier, 1979 and Giggenbach, 1988). Other similar geothermometers also take into account the concentration of magnesium (Giggenbach, 1988) and lithium (Fouillac and Michard, 1981). These "classical" geothermometers and several of their modifications have been successfully applied to many geothermal waters and can be easily implemented through simple equations either directly or through various available software packages (e.g., Verma et al., 2008 and Powell and Cumming, 2010). As a result, classical geothermometers have become important and essential geothermal exploration tools. However, these geothermometers can fail because of the assumptions on which they are based on. The Na–K geothermometers assume chemical equilibrium between the fluid and the minerals albite and K-feldspar, while the K–Mg geothermometers consider equilibrium of the fluid with muscovite, clinochlore and K-felspar (Giggenbach, 1988). These minerals are usually common in geothermal systems, however their composition may vary. Also, especially in lower temperature systems, the Na–K and K–Mg ratios might be controlled by other minerals such as smectites, causing these geothermometers to fail. In addition, geothermal fluids ascending to ground surface are typically affected by gas loss, mixing and/or dilution with shallower waters, masking their deep geochemical signatures (Fournier, 1977).

In the late 1970s and early 1980s, with the availability of increasingly powerful computers, numerical multicomponent geochemical models were developed specifically for the study of hydrothermal systems, with direct application to chemical geothermometry (e.g., Michard et al., 1981, Michard and Roekens, 1983, Arnorsson et al., 1982, Arnorsson et al., 1983a, Arnorsson et al., 1983b, Reed, 1982 and Reed and Spycher, 1984). These studies showed that given a fluid composition, numerical models could be used to compute the (theoretical) equilibration temperature of a suite of reservoir minerals and thus infer reservoir temperature. Such multicomponent approaches present advantages over classical geothermometers because they rely on complete fluid analyses and a solid thermodynamic basis, rather than the solubility of a few minerals or (semi-)empirical correlations, and thus in principle apply to any geochemical system. However, because multicomponent geothermometry methods require a numerical model and a priori assumptions regarding reservoir minerals, these methods are much less practical than classical geothermometers. For this reason, multicomponent geothermometry methods have been applied much less frequently than classical geothermometers.

Here, we revisit the multicomponent chemical geothermometry method presented by Reed and Spycher (1984) and further developed by Pang and Reed (1998) and Palandri and Reed (2001).

The method consists of using full chemical analyses of water samples to compute the saturation indices ($\log(Q/K)$) of reservoir minerals over a range of temperatures (e.g., 25–300 °C). The saturation indices are graphed as a function of temperature, and the clustering of $\log(Q/K)$ curves near zero at any specific temperature (for a group of certain reservoir alteration minerals) is inferred to yield the reservoir temperature (e.g., Fig. 1a). Prior to computing saturation indices, the composition of the deep fluid is reconstructed by applying corrections for any dilution and/or mixing effects and by numerically adding back (to the analyzed fluid composition) any gases that may have exsolved from the deep fluid on its way to the ground surface. As shown by these earlier studies, the scatter of $\log(Q/K)$ curves can also be used to decipher, and correct for, dilution and/or other processes affecting the evolution of deep geothermal fluids.

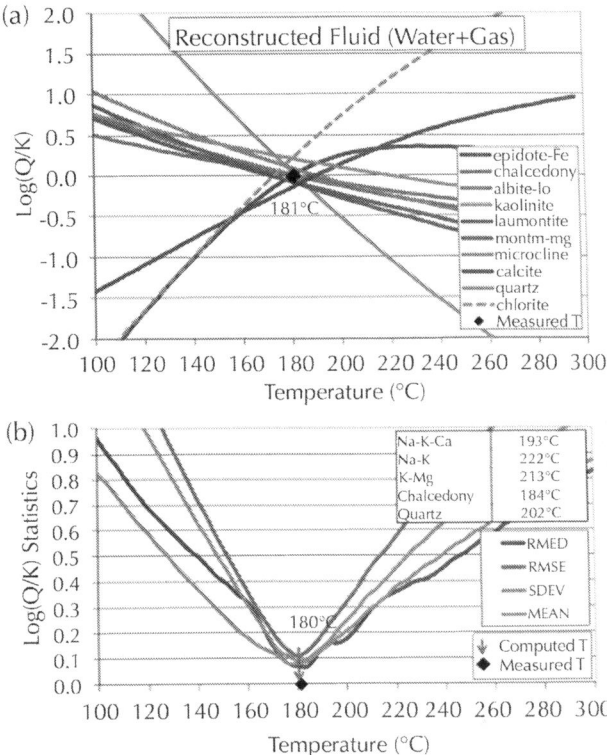

Figure 1: Multicomponent geothermometry using an Icelandic geothermal water (Hveragerdi Well 4, Arnorsson et al., 1983a) previously used as an example by Reed and Spycher (1984): (a) Computed saturation indices, log(Q/K), as a function of temperature, showing clustering near zero close to the measured temperature (181 °C). The amount of $H_2O_{(g)}$ in steam was adjusted by numerical optimization (see text). (b) Statistical analyses of saturation indices: median (RMED), mean root square error (RMSE), standard deviation (SDEV) and average (MEAN) of absolute log(Q/K) values. The reservoir temperature is inferred from the temperature at which RMED is minimum (see text). Results of classical geothermometers are also shown for comparison, calculated using the reconstituted deep fluid composition (Na–K–Ca, Fournier and Truesdell, 1973; Na–K, Giggenbach, 1988; chalcedony, Fournier, 1977; quartz, Fournier and Potter, 1982).

The method of Reed and Spycher (1984) was developed for single-point fluid analyses and, although quite powerful, requires somewhat tedious data processing followed by temperature estimations relying on fairly subjective "eyeballing" of the clustering of computed $log(Q/K)$ curves. For these reasons, the first goal of this study was to implement the method into a practical software tool that uses a set of objective criteria to estimate reservoir temperatures from computed mineral saturation indices for any given input water composition. The second goal was to integrate the method with numerical optimization to allow simultaneous processing of multiple water analyses to estimate reservoir temperatures as well as unknown input parameters affecting the temperature estimations (such as the amount of dilution, if any, the degassed steam fraction and/or composition, or the concentrations of aqueous species that may not have been measured or that do not reflect reservoir conditions). The approach followed in this study is essentially the automation of the method presented by Reed and Spycher (1984) into a stand-alone computer program (GeoT²), to ease and standardize temperature estimations, and to allow numerical optimization of model input parameters using existing parameter-estimation software. This paper presents the methodology implemented into the GeoT computer program, and its application to two previously published examples of geothermal waters. The program is then applied to experimental and simulated fluid compositions to examine the impact of fluid–rock disequilibrium on estimated temperatures by both multicomponent and classical geothermometry. This work extends preliminary materials presented earlier in a conference paper (Spycher et al., 2011). In a complementary paper (Peiffer et al., 2014), the integrated geothermometry approach presented below is applied to the Dixie Valley geothermal system, Nevada, including the simultaneous processing of multiple datasets with numerical optimization to provide insights on this complex geothermal field. The pros and cons of multicomponent and classical geothermometry is also further investigated by Peiffer et al. (2014) andWanner et al., 2013 and Wanner et al., 2014, with application of both methods to Dixie Valley.

TECHNICAL APPROACH

A computer program (GeoT, Fig. 2) was developed drawing on existing routines and methods implemented into programs TOUGHREACT (Xu et al., 2006 and Xu et al., 2011), SOLVEQ/CHILLER (Reed, 1982 and Reed, 1998) and GEOCAL (Spycher and Reed, unpublished, U. Oregon, 1985). The core of the software is essentially a homogeneous geochemical speciation algorithm solving mass-balance/mass-action equations by Newton–Raphson iterations (e.g., Reed, 1982). Using complete fluid analyses, the saturation indices of minerals ($\log(Q/K)$) are obtained from the computed ion activity product (Q) and thermodynamic equilibrium constant (K) for each mineral. Data for the computation of activity coefficients together with equilibrium constants for aqueous complexes and minerals at various temperatures are read from an external thermodynamic database. The pH at elevated temperatures is calculated from the total numerical H^+ concentration computed from the input (known) low-temperature pH, following the method ofReed and Spycher (1984). The program allows for simultaneous regression of multiple waters, automatic reconstitution of deep fluid compositions, and estimation of reservoir temperature by numerically assessing the clustering of computed mineral saturation indices, as described below.

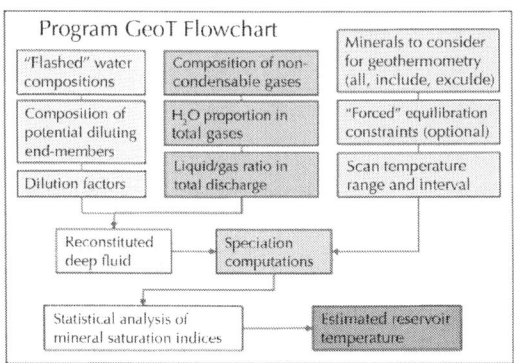

Figure 2: General structure of the multicomponent geothermometry computer program GeoT. See Fig. 3 for the temperature estimation methodology.

Deep Fluid Reconstruction

Input water compositions are corrected for gas loss, dilution or concentration (by some factor), and/or mixing with other waters. If known, the gas phase composition and its proportion in the fluid, the dilution/concentration factor, and/or compositions and fractions of mixing water end-members are entered into the calculations (Fig. 2). Following Reed and Spycher (1984), gases are numerically added back to the fluid in their respective stoichiometric forms (as defined in the input thermodynamic database; e.g., H_2S as $1\times HS^-$ and $1\times H^+$; CO_2 as $-1\times H_2O$, $1\times HCO_3^-$ and $1\times H^+$), after calculation of the total numerical H^+ of the solution from a given pH. The new total numerical H^+ of the solution, after gas addition, is then used to compute the pH of the reconstructed fluid. The "de-mixing" of a particular end-member solution is carried out essentially in the same manner, by removing or adding a particular component in given input stoichiometric proportions.

Unknown concentrations of one or more specific elements in the deep fluid can be optionally computed by assuming that the concentration of each element is constrained by thermodynamic equilibrium between that element and a respective mineral. This follows the approach presented by Pang and Reed (1998) for aluminum ("Fix-Al" method), and Palandri and Reed (2001) for silica. This equilibrium constraint can be applied over the entire scanned temperature range, only at the initial speciation temperature, or in both instances, as chosen for each dissolved species.

Another approach developed here to help the reconstruction of deep fluid compositions is the estimation of unknown input concentrations (e.g., Al, Mg) or parameters (e.g., dilution factor, gas amount) by numerical optimization of the clustering of mineral saturation indices near zero. This approach is discussed further below and in Peiffer et al. (2014).

Temperature Estimation from Clustering of Saturation Indices

Several schemes were evaluated to estimate reservoir temperatures based on the saturation indices of a group of minerals. Best matches of known temperatures (when reported) were obtained by taking the temperature at which the absolute value of the median of computed saturation indices (for a group of minerals) is minimum, after applying an optional elimination algorithm to select a certain number (N_{best}) of "best-clustering" minerals (Fig. 3). This is because this non-parametric value ensures that large saturation indices of minerals falling significantly off-cluster do not bias computed temperatures. The temperature estimation can be based on a given suite of specific minerals, or applied "blindly" using all minerals available in a separate thermodynamic database. The number of "best-clustering" minerals can be specified (e.g., of a suite of fifteen input minerals, only $N_{best} < 15$ minerals are used for the final temperature determination) or defaults to the total number of considered minerals. The program first eliminates all minerals whose absolute value of saturation indices remain at or above 0.05 (an arbitrary but near-zero cut-off value) over the entire temperature range considered. For remaining minerals, the median of the absolute values of saturation indices (RMED) is then evaluated at all temperatures. Optionally (and recommended), the list of minerals contributing to the temperature estimation is further narrowed down by eliminating minerals with saturation indices exceeding RMED × 1.2 (an arbitrary value > 1 tested to yield satisfactory results). This is done in an iterative manner during which RMED is continuously re-evaluated (Fig. 3). The temperature at which the final RMED value is at a minimum (T_{RMED}) is then inferred to be the reservoir temperature. The temperatures at which other statistical parameters are at a minimum (mean, standard deviation, and mean-root-square error) are also computed to provide information on the quality of the clustering (e.g., Fig. 1b). In the case of perfect clustering, these other statistical parameters become zero at the same temperature, and therefore their values can be used (in addition to the magnitude of RMED) to further assess the degree of confidence in computed T_{RMED} values.

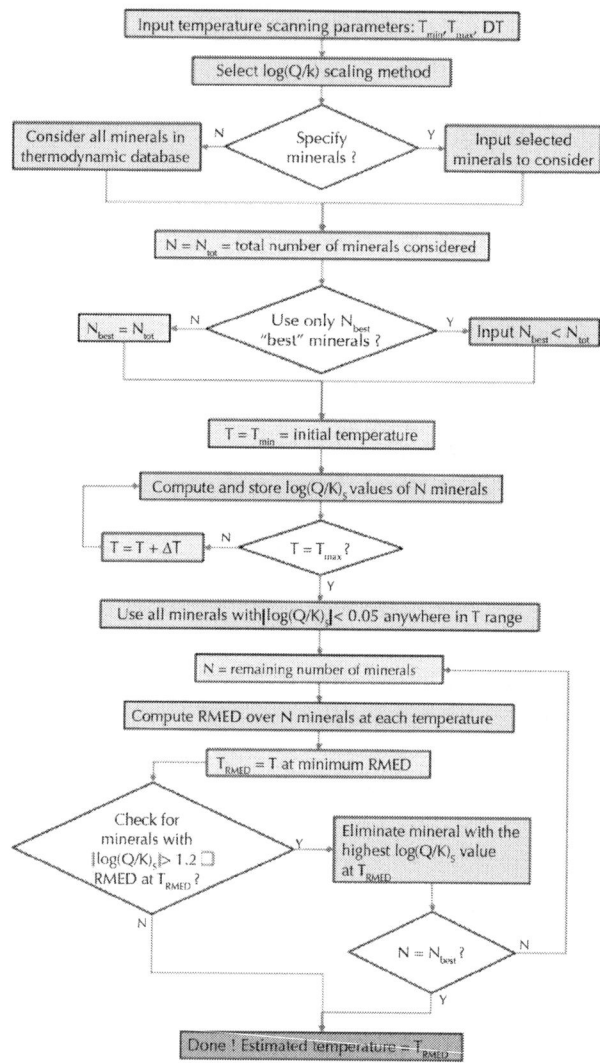

Figure 3: Flowchart of the scheme implemented in this study to estimate reservoir temperatures from computed mineral saturation indices. These indices are optionally scaled ($\log(Q/K)_s$). RMED is the median of absolute values of saturation indices (median of $|\log(Q/K)_s|$) (see text).

Mineral saturation indices are a direct expression of reaction thermodynamic affinity (Gibbs free energy of reaction), and therefore are extensive variables that are function of the mineral formula. The temperature at which $\log(Q/K) = 0$ (equilibrium point) for a given mineral is not a function of the mineral formula, but the $\log(Q/K)$ values away from this equilibrium point (more precisely, the slopes of the $\log(Q/K)$ values with temperature around the equilibrium point, e.g., Fig. 1a) are function of the mineral formula. For example, defining a clay or a zeolite mineral on the basis of a different total number of silica or oxygen atoms (as is often the case) yields different $\log(Q/K) \neq 0$ values, which in some cases can affect the mineral selection and minimization procedure described above. For this reason GeoT implements two optional $\log(Q/K)$ scaling approaches. The first one consists of dividing $\log(Q/K)$ values by the molecular weight of the mineral. The second (preferred) method is applied only to silicate minerals, and consists of dividing $\log(Q/K)$ values by the number of silicon atoms in the mineral formula. These schemes are similar in nature to the scaling approach implemented by Wolery (1979) and Reed, 1982 and Reed, 1998, which consists of dividing $\log(Q/K)$ values by the sum of absolute values of the stoichiometric coefficients of species in the reaction expressing mineral dissolution. The advantage of our approach is that the scaling depends only on mineral unit formula and not on the type of species used to describe a mineral (e.g., as when calcite is expressed as a function of Ca^{+2} and CO_3^{-2} versus H^+, HCO_3^-, and Ca^{+2}).

Numerical Optimization

If unknown, input parameters necessary to reconstruct the deep fluid composition can be estimated by adjusting these parameters to minimize the cluster of mineral saturation indices near zero. This approach was shown to work well to "un-dilute" and re-carbonate geothermal waters (Reed and Spycher, 1984 and Pang and Reed, 1998) and oil field brines (Palandri and Reed, 2001) affected by mixing and degassing. Here, because the calculations to reconstruct the deep fluid, and the geothermometry computations, are performed by one stand-alone program (Fig. 1), optimization of the clustering can be carried out numerically using external software (e.g., PEST, Doherty, 2008; iTOUGH2, Finsterle and Zhang, 2011). This allows a larger number of parameters to be estimated than manually by trial-and-error, and can be accomplished for several water compositions simultaneously if these waters are all believed to originate from one common geothermal reservoir (e.g., Peiffer et al., 2014). In principle, any input parameter could be estimated by numerical optimization; however the optimization may not be successful if too many input parameters were to be estimated simultaneously.

The program computes various statistical parameters that can be used to formulate objective functions for numerical optimization. Main output parameters include the median, mean, standard deviation and root mean-square error of saturation indices as a function of temperature (e.g., Fig. 1b), and the temperatures at the minimum value of these statistical parameters (T_{RMED}, T_{SDEV}, T_{MEAN}, T_{RMSE}, respectively). Several other clustering measures are computed, including the spread of temperatures (DT) given by the points at which $\log(Q/K) = 0$ for each individual mineral, and the average and standard deviation of these temperatures (T_{DT} and σ_{DT}, respectively). For a perfectly clustered system, T_{RMED}, T_{SDEV}, T_{MEAN}, T_{RMSE} and T_{DT} should all be identical, and DT and σ_{DT} should both equal zero. Therefore, when processing each individual water analysis, these outputs can be used directly for numerical optimization, by simultaneously: (1) taking T_{RMED} as the estimated temperature, (2) minimizing DT or σ_{DT}, and (3) minimizing the difference between T_{RMED} and the average of T_{SDEV}, T_{MEAN}, T_{RMSE}. When processing more than one water analysis, the mean and standard deviation of T_{RMED} and DT values computed for all the different water compositions are also output to provide additional parameters for use in a global optimization.

Numerical optimization using the above-described simple statistical parameters worked reasonably well for all the cases tested (see examples below and Peiffer et al., 2014). However, it should be noted that successful optimizations typically required testing various search procedures (as implemented into the optimization software) and initial parameter guesses. In this study, the simplex method (derivative-free) was used and tested to produce results similar to those obtained with a much slower but more fail-proof grid-based approach. The specifics of these various numerical optimization methods can be found elsewhere (e.g., Finsterle and Zhang, 2011 and references therein) and for this reason are not discussed here.

THERMODYNAMIC EQUILIBRIUM AND THERMODYNAMIC DATA

As in the case of classical solute geothermometry, the main underlying assumption of multicomponent geothermometry is that chemical equilibrium (or near-equilibrium) exists between the deep geothermal fluid and certain minerals in the reservoir. The degree to which thermodynamic equilibrium is reached in different geothermal reservoirs certainly varies, although there is consensus from many studies that local (partial) equilibrium prevails between the fluid and alteration minerals in most geothermal reservoirs (e.g.,Reed, 1997 and references therein). The most compelling evidence for equilibrium is that classical and multicomponent geothermometers have been shown to work in numerous instances. Also, thermodynamic equilibrium models have been successfully applied to reproduce observed mineral assemblages in many hydrothermal systems.

The degree to which disequilibrium affects solute geothermometry is not clear-cut and has not been evaluated in a quantitative manner. For this reason, our study also aims at examining the extent to which the approach described earlier can be used with systems that deviate from thermodynamic equilibrium. This is done by applying classical and multicomponent geothermometry to fluids obtained from short-term water–rock reactions experiments at elevated temperatures, and to synthetic fluids sampled from numerical simulations of water–rock interactions at various degrees of disequilibrium. Further work using numerical reactive transport models to evaluate the effect of a geothermal fluid's chemical evolution (from depth to ground surface) on solute geothermometry is presented by Peiffer et al. (2014) and Wanner et al., 2013 and Wanner et al., 2014.

Because multicomponent geothermometry is based on thermodynamic equilibrium computations, the sensitivity of the method to differences in input thermodynamic data also needs to be evaluated. In this study, the thermodynamic database compiled by Reed and Palandri (2006) (soltherm.h06) was selected, because this database was developed primarily for high temperature applications (to 350 °C along the saturation pressure of water) and was found to provide satisfactory results. This database relies on Gibbs free energy data primarily from Holland and Powell (1998) for minerals, and primarily from SUPCRT92 (Johnson et al., 1992 and Shock et al., 1997) for aqueous species. Effects of alternative sources of thermodynamic data on temperature predictions were previously investigated (Spycher et al., 2011) using alternative databases from Sandia National Laboratories (2007) (data0.ymp) and Blanc et al., 2007 and Blanc et al., 2012 (thermoddem). Results indicated that the choice of thermodynamic data can affect temperature predictions. This previous work, as well as investigations carried out for the present study, showed reasonable agreement in most cases when using these alternative databases. However, deviations up to 20 °C in estimated temperatures were observed in other cases, mostly because of differences in the solubility constants (and composition) of certain minerals, but also because of differences in aqueous species dissociation constants (as, for example, with aluminum species). This does not imply that these alternative thermodynamic databases are not as reliable as the one selected here. It simply points out the importance of continued development and evaluation of thermodynamic data for applications to chemical geothermometry, particularly at elevated temperatures. Some databases may perform better with some species/minerals or over certain temperature ranges. However, a detailed evaluation of thermodynamic data included in these databases was not within the scope of this study.

RESULTS

We present below examples of applications of our approach. Previously mentioned studies have already documented the merits of multicomponent geothermometry and its application to a wide range of natural thermal waters. Here the focus is given to testing temperatures computed with the newly developed geothermometry code (Fig. 2 and Fig. 3), in some cases coupled with numerical optimization using iTOUGH2 (Finsterle and Zhang, 2011). First, an example from a previous study using an Icelandic water is revisited (Fig. 1). A second example using a geothermal fluid from Long Valley, California, is then presented (Fig. 4). We then also examine the effect of thermodynamic dis-equilibrium on estimated temperatures using waters obtained from water–rock interaction experiments (Fig. 5) and reactive transport simulations (Fig. 6). For all discussed examples, computed temperatures are also compared with results of classical geothermometers.

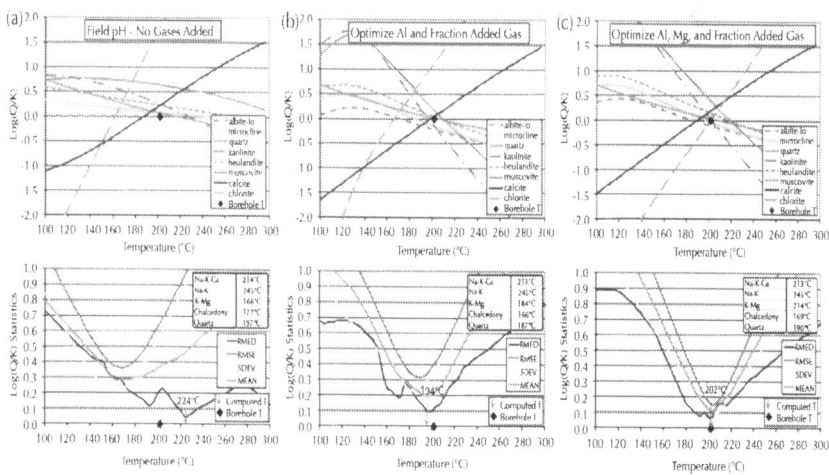

Figure 4: Multicomponent geothermometry using a geothermal water from Long Valley (Shady Rest area, Farrar et al., 1989). (a) Results using

the original fluid composition and field pH without added gases, fixing the Al concentration by equilibrium with albite; (b) Results after correcting for gas loss and optimizing the fraction of gas lost and the Al concentration by numerical optimization; (c) same as (b) with numerical optimization of the Mg concentration. See text. The temperature is determined from the saturation indices of all minerals shown. Results of classical geothermometers are also shown for comparison, calculated using the fluid composition before (a) and after (b, c) optimization (see caption of Fig. 1 for references on these geothermometers).

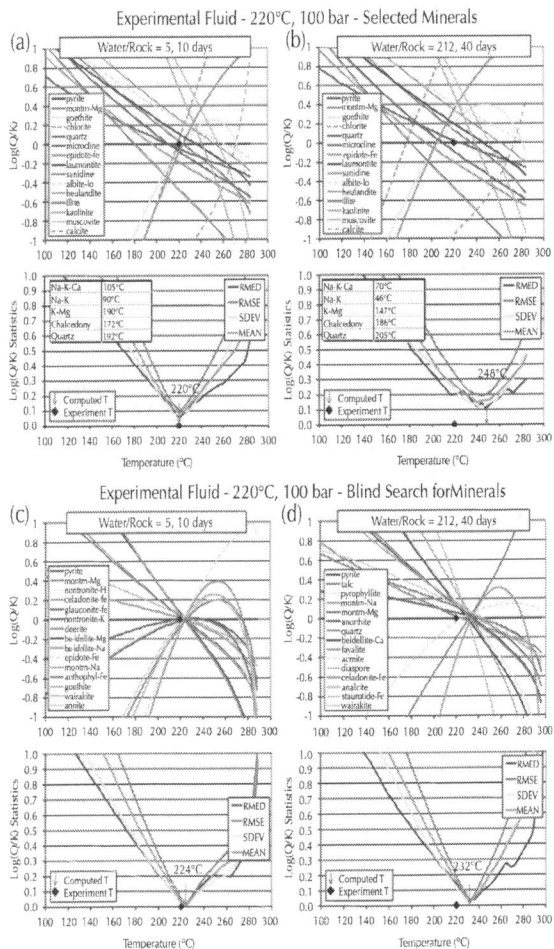

Figure 5: Multicomponent geothermometry applied to fluids obtained after reacting volcanic tuff with a 0.1 m NaCl solution at 220 °C, 100

bar, at two different water/rock (weight) ratios and reaction times, in two separate experiments (left and right plots, respectively) using the same rock material (powdered). Top graphs (a and b): results showing GeoT temperature estimations using the ten best-clustering minerals (top ten listed) out of 15 selected input minerals. Bottom graphs (c and d): results showing GeoT temperature estimations using the 15 best-clustering minerals out of all minerals available in the input thermodynamic database, with (c) and (d) showing examples of good and erroneous results, respectively (many minerals are not reasonable candidates).

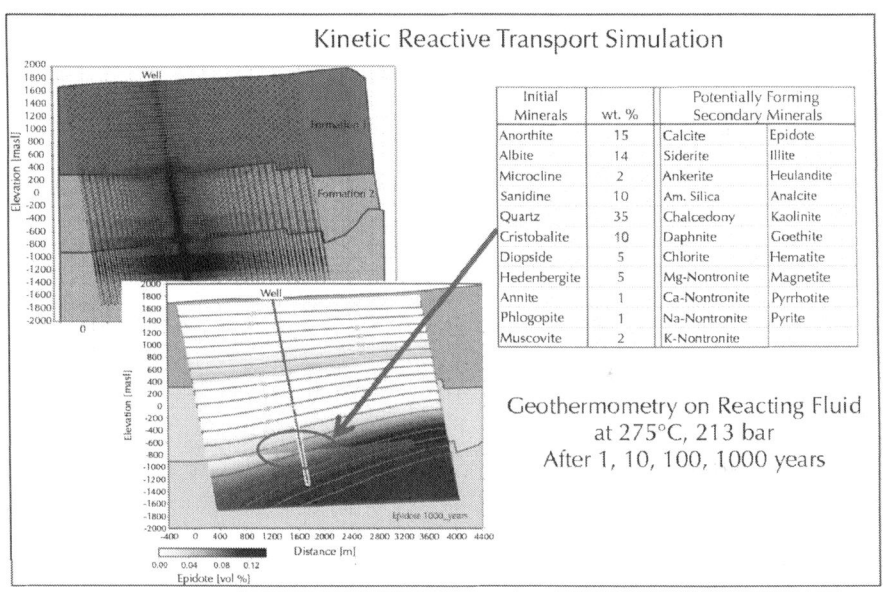

Figure 6: Testing multicomponent geothermometry with results of reactive transport simulations of an Enhanced Geothermal System (Sonnenthal et al., 2012). Plots on the left show the reactive transport model numerical grid (top) and results including computed isotherms, the predicted distribution of epidote, a typical reservoir alteration mineral (bottom), and the observed epidote distribution (green shading in well below about the 200 °C isotherm). The table on the right shows the initial mineralogy in the model grid block selected for geothermometry (indicated by arrow), and the secondary minerals allowed to form (kinetically) at that location. The simulated evolution of the mineralogy and fluid composition at this location is shown in Fig. 7.

Example Applications to Geothermal Waters

The first example is taken from Reed and Spycher (1984) (Fig. 1). The deep temperature of a geothermal fluid is computed using water and gas analyses from Arnorsson et al. (1983a) for a geothermal well in Iceland. In the present case, fifteen minerals are given as potential reservoir minerals, and the deep fluid temperature is computed from the ten "best" clustering minerals (those shown in Fig. 1, automatically selected following the procedure shown in Fig. 3; the other minerals considered are goethite, illite, muscovite, anhydrite and fluorite). The temperature is scanned in steps of 2 degrees. In absence of measurements, the steam weight fraction in the total (water + steam) discharge is initially estimated (~0.04) using the enthalpy of pure water at the measured downhole fluid inflow temperature (181 °C) and the enthalpy of pure water and steam at the sampling pressure, assuming adiabatic flashing (Arnorsson et al., 1983a). The amount of H_2O gas in the steam (0.9992 mole fraction) is calculated using the ideal gas law and the reported proportions of gas and condensate (1.06 l gas per kg condensate). This results in a computed temperature (T_{RMED}) of 172 °C and an average temperature ($T_{DT} \pm \sigma_{DT}$) of 180 ± 17 °C. The temperature calculation can be refined by numerical optimization to estimate either the steam weight fraction (yielding ~0.06) or the mole fraction of H_2O gas in the steam (yielding ~0.9986), with essentially identical results, the latter yielding a computed temperature (T_{RMED}) of 180 °C and an average temperature ($T_{DT} \pm \sigma_{DT}$) of 186 ± 14 °C (Fig. 1). Obviously, results are also a direct function of the selected mineral assemblage (and their thermodynamic data). The results of classical geothermometers for this case vary (Fig. 1b), with chalcedony indicating a temperature consistent with the measured value of 181 °C, and the Na–K geothermometer showing the largest deviation (about 40 °C higher).

The second example consists of a geothermal water from Long Valley, California. The fluid analysis is that reported by Farrar et al. (1989), for a sample collected with a bailer by the USGS from a geothermal well at the Shady Rest area, where the reported maximum wellbore temperature is 202 °C. This example was chosen to illustrate the reconstruction of a deep fluid after loss of gases, including numerical optimization of unknown or poorly constrained input parameters. In absence of reported gas analyses for this particular well, the gas composition used to reconstitute the deep fluid are taken as the average of analyses reported by Farrar et al. (1985) for springs and fumaroles in the Long Valley, Casa Diablo area. In this example, reservoir temperatures are estimated using nine minerals (quartz, microcline, albite, chlorite, muscovite, heulandite, calcite, kaolinite, and pyrite) selected on the basis of available petrologic data (e.g., Tempel et al., 2011) and preliminary analyses considering a larger number of minerals. Three cases of temperature estimation are tested. In the first case (Fig. 4a), the original water composition is used, without gases added, but taking the reported field pH value (5.9; laboratory pH not reported), thus presumably reflecting minimal degassing. In this case, the total dissolved carbonate content is computed by charge balance, thus correcting for some loss of CO_2, and the Al concentration is fixed by assuming equilibrium of the fluid with albite at all temperatures (the "Fix-Al" method of Pang and Reed, 1998). In the second case (Fig. 4b), gases are added back into solution, after taking the reported laboratory measurement of alkalinity for total dissolved carbonate, and computing the laboratory (degassed) pH from charge balance (7.3 at 25 °C). In this case, the gas fraction (0.12) and the Al concentration (~0.5 ppm) are estimated by numerical optimization using GeoT with iTOUGH2, and the pH at temperatures below boiling (in the range 25–100 °C) is computed to be around 5.6, slightly lower than the field pH measurement. The third case (Fig. 4c) is essentially the same as the second case, except that numerical optimization is used to estimate the Mg concentration (~0.02 ppm) in addition to the gas fraction (0.093) and Al concentration (~0.15 ppm).

These three cases (a, b, and c) result in temperature estimates that are progressively better constrained and closer to the maximum reported wellbore temperature of 202 °C (Fig. 4): reservoir temperatures values (T_{RMED}) of 224 °C, 194 °C, and 202

°C, respectively, and average temperatures values ($T_{DT} \pm \sigma_{DT}$) of 219 ± 39 °C, 194 ± 16 °C, and 204 ± 12 °C. The fact that T_{RMED} exactly matches the maximum reported wellbore temperature in case c) is likely fortuitous, as seen by the spread in estimated temperatures ($T_{DT} \pm \sigma_{DT}$). In all three cases, the temperatures estimated by the Na–K–Ca (213–214 °C) and quartz (187–197 °C) geothermometers fall the closest to the reported borehole temperature, and mostly within the range of (but higher than) GeoT estimates (Fig. 4). The K–Mg geothermometer performs poorly when the measured Mg concentration is used, but yields reasonable results consistent with the Na–K–Ca geothermometer when the deep-fluid Mg concentration is estimated by numerical optimization (Fig. 4c).

Method Testing and Application Using Experimental Data

As part of another ongoing study (Saldi et al., in preparation), volcanic tuff samples from the Desert Peak geothermal area, Nevada, were reacted with a 0.1 m NaCl solution under conditions of constant temperature and pressure and at various water/rock ratios (Appendix A). Here, we compare results of classical and multicomponent geothermometry with fluid samples collected from relatively short-term (10 and 40 days) experiments at 220 °C and 100 bar (Fig. 5), with the goal of examining the effect of disequilibrium on temperature predictions. Two cases are considered: a 10-day experiment at low water/rock weight ratio (~5) (Fig. 5a and c) and a 40-day experiment at high water/rock weight ratio (~212) (Fig. 5b and d), thus with a smaller total surface area available for reaction than in the case of the first experiment. Both time and water/rock ratio (available reactive surface area) dictate the extent of reaction. Therefore, in the first experiment, compared to the second experiment, the effect of ~40× lower water/rock ratio is greater than the 4× shorter reaction time, yielding fluids that have reacted more fully and that are potentially closer to equilibrium with the reacted rock than in the second experiment. From bulk XRD measurements of samples taken prior to the experiments, the rock is known to contain quartz (~37 wt.%), K-spar (~40 wt.%, reported as sanidine), kaolinite (~15 wt.%) and calcite (~4 wt.%), the remaining 4% being either amorphous or below XRD detection limits (typically ~1 wt.%). In a first computation, these minerals and a suite of other potentially relevant minerals are specified (fifteen altogether, Fig. 5a and b). Temperature calculations are set to use the ten best-clustering of these minerals. Temperature increments of 4 °C are specified. Redox is constrained by computing unknown sulfide concentrations (presence detected by odor) using the measured Fe and sulfate concentrations and assuming that pyrite remains at equilibrium with the solution at all temperatures. Using these inputs, the case with a low water–rock ratio yields a correct temperature (T_{RMED} = 220 °C, T_{DT} = 224 °C ± 19 °C as $_{DT}$) (Fig. 5a), even though the system is not close to equilibrium with all selected minerals. In contrast, classical geothermometers fail, significantly underpredicting the experiment temperature, with the closest result given by the quartz geothermometer (192 °C, Fig. 5a). This is a direct result of the lack of equilibrium after short reaction times, compounded by the introduction of 0.1 m NaCl in the reacting solution. It is also observed that the temperature given by the quartz saturation index (210 °C) is closer to the measured temperature than the equation-based quartz geothermometer. The case with a high water–rock ratio shows even further departure from equilibrium (Fig. 5b). In this case, our multicomponent geothermometry approach significantly overestimates the temperature (T_{RMED} = 248 °C, T_{DT} = 241 °C ± 24 °C as $_{DT}$), but still yields smaller deviations from the actual temperature than most of the classical geothermometers (Fig. 5b).

Obviously, the number and types of minerals considered can significantly affect temperature estimations. In the low water–rock ratio case (Fig. 5a), reducing the number of minerals from which the temperature is determined from ten to six (using the same initial list of fifteen minerals) does not significantly affect the computed temperature (T_{RMED} = 224 °C). However, when specifying only the five minerals known to be present in the original rock sample (quartz, calcite, kaolinite, and sanidine), the temperature cannot be correctly estimated (T_{RMED} = 200 °C and T_{DT} = 231 °C ± 34 °C as σ_{DT}). This points to the importance of considering as many secondary (alteration) minerals as possible when known. Also, including minerals that display both prograde and retrograde solubility behavior significantly helps constrain the temperature determinations.

When the mineralogy is poorly known, "blind" searches can be conducted for a large number of minerals. However, this can yield fairly unpredictable results and should be used with caution, only as a first step to evaluate a certain geochemical system before narrowing down the list of minerals. As an example, the experimental fluids described above were processed using a "blind" search for fifteen best-clustering minerals (out of about 200 possible minerals in the input thermodynamic database). Although this yielded reasonable results for the case of low water–rock ratio (Fig. 5c), a larger discrepancy is obtained for the case of higher water–rock ratio (Fig. 5d). In both cases and particularly the latter, many minerals were selected that are not reasonable candidates on the basis of kinetic grounds (i.e., slow-forming high-temperature/ pressure metamorphic minerals).

Predictions for Simulated Waters under Partial Equilibrium Conditions

In this concluding example, multicomponent geothermometry computations are applied to synthetic waters obtained from 2-D reactive transport simulations of a cross-section through the western flank of Newberry Volcano, a Department of Energy Enhanced Geothermal System Demonstration Project in the Western United States (Sonnenthal et al., 2012) (Fig. 6). In this simulation, groundwater recharge (with a chemical composition measured at the site) provides the fluid which penetrates to a depth of over 3 km where it heats up and reacts with primarily rhyolitic tuffs and shallow granodioritic intrusive rocks, with the mineralogical abundances shown in Fig. 6 (note that plagioclase is approximated with a mixture of albite and anorthite endmembers). The simulated steady-state temperature and fluid pressure distributions were modeled by calibrating thermal conductivities and temperatures to data from deep geothermal exploration wells. The starting mineral assemblage in the simulation consists solely of the primary igneous minerals, with the objective of comparing model results to observed secondary minerals in deep boreholes. Reactive transport is simulated for several thousand years under kinetic constraints using a new parallelized version of TOUGHREACT V2 (Xu et al., 2011) and kinetic data from the literature. The same thermodynamic database (Reed and Palandri, 2006) is used as with the GeoT computations, ignoring effects of pressure above the pure water saturation pressure.

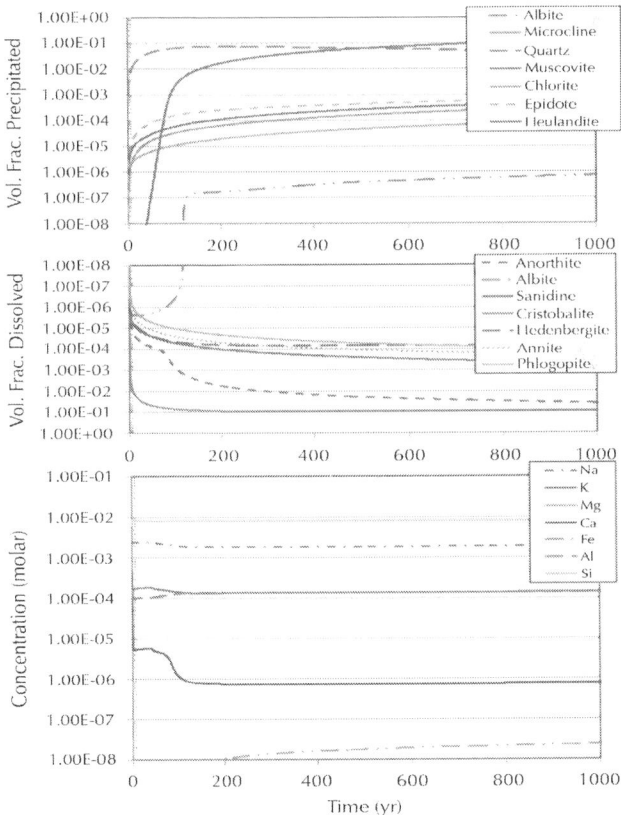

Figure 7: Computed mineral assemblage (top) and fluid composition (bottom) in a model grid block at 275 °C and 213 bar. The simulated fluid

composition after 1000 years remains nearly constant. Compositions at 1, 10, 100, and 1000 years are used as input to the geothermometry computations shown in Fig. 8.

Geothermometry is applied to fluid sampled in a model grid block at 275 °C (Fig. 6) after simulated times of 1, 10, 100, and 1000 years, to capture a range of compositions reflecting chemical evolution toward local equilibrium. In this simulation, the primary rock minerals react with recharging water, essentially resulting at depth in the precipitation of quartz, zeolites, epidote, chlorite, micas and feldspars, similar to the observed alteration mineral assemblage. The hot fluid reaches a nearly steady chemical composition after a few hundred years (Fig. 7).

The reservoir temperature is computed using the entire list of minerals shown in Fig. 6, with the final temperature determined from the six best-clustering minerals (Fig. 8). The temperature is scanned in steps of 2 °C. After only one year of reaction, the computed temperature is somewhat over-predicted (T_{RMED} = 286 °C and T_{DT} = 269 °C ± 41 °C as σ_{DT}) but closer than estimates using classical geothermometers (the quartz geothermometer coming the closest at 255 °C, Fig. 8a). The fluid after 1000 years yields a reasonable temperature estimate (T_{RMED} = 278 °C and T_{DT} = 282 °C ± 9 °C as σ_{DT}), but classical geothermometers still deviate significantly from the simulated temperature (with Na–K–Ca and quartz coming closest at 256 °C).

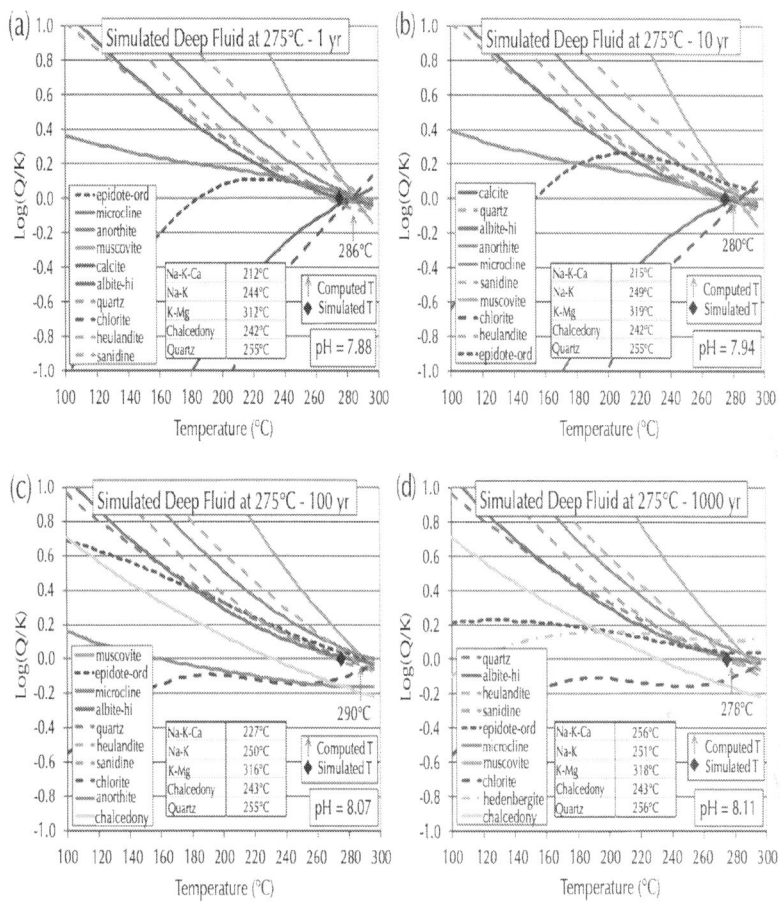

Figure 8: Results of multicomponent and classical geothermometry applied to a simulated fluid at 275 °C (Fig. 6 and Fig. 7) after reaction times of (a) 1 year, (b) 10 years, (c) 100 years and (d) 1000 years. The temperature is computed given the list of all minerals shown in Fig. 6, and keeping the six best-clustering minerals. The minerals in the legend of each graph are listed in order of increasing departure from equilibrium (from the top of the list) at the computed temperature shown (temperature is determined from the six top minerals listed for each graph).

It should be noted that as time evolves, different minerals are picked (by the program) for temperature estimations, as these minerals come closer to equilibrium with the solution. In the case of the 1-year fluid, the best clustering is determined to occur with epidote, microcline, anorthite, muscovite, calcite and albite. After longer simulation times, anorthite is predicted to fully dissolve (as would be expected). Also, the fluid saturation with respect to calcite (initially at saturation but not present in the reacting rock) decreases with time. Consequently, with longer reaction times, the list of minerals picked for temperature estimations changes, and includes quartz, albite, heulandite, sanidine, epidote and microcline at 1000 years (Fig. 8d). It should also be noted that the majority of the minerals picked for temperature computations are those that are forming during reaction (Fig. 7, top graph). Including more minerals in the temperature estimation (e.g., all precipitating and dissolving minerals shown in Fig. 7) somewhat increases the predicted temperature (T_{RMED} = 286 °C at 1000 years) but considerably expands the range of computed temperatures (T_{DT} = 240 °C ± 67 °C as $_{DT}$), because many of the dissolving minerals remain below saturation levels. This reinforces conclusions from previous studies that this geothermometry method is most successful when used with currently forming alteration minerals, rather than primary minerals, which can be a limitation of the method because this information may not always be known.

CONCLUSIONS

A solute chemical geothermometry approach was developed by automating the multicomponent chemical geothermometry method of Reed and Spycher (1984) into a new stand-alone computer program (GeoT) and integrating this program with parameter estimation software. Without optimization, GeoT is useful for fast geothermometry computations for fluids from single locations, and for independently processing fluids from multiple locations in one single program run. Numerical optimization considerably extends the power of this software because it allows the estimation of unknown or poorly known parameters that are needed to compute reservoir temperatures. This includes, for example, estimating the proportion and/or chemical composition of gases lost before sampling, or the concentrations of some elements such as Al and Mg which are typically poorly constrained (see Peiffer et al., 2014). Such optimization can be performed for single water analyses (as presented here) or by processing simultaneously multiple analyses of various waters presumed to originate from a single reservoir (as done by Peiffer et al., 2014). However, numerical optimization remains a challenge for systems with highly non-linear behavior and multiple local minima in the objective function, often requiring grid-based search procedures that are not as computationally efficient as other methods. In this study, the focus was given on presenting the concept of numerical optimization in the context of multicomponent geothermometry, without delving into the complex subject of numerical optimization itself. It is clear that more room remains to explore this subject, including the derivation and testing of additional objective functions as well as minimization procedures that can provide computationally efficient optimization alternatives specifically geared toward maximizing the clustering of mineral saturation indices.

Finally, it must be made clear that the approach developed in this study is not intended to replace classical geothermometry, which remains an important and essential tool in geothermal exploration. However, in cases when classical geothermometers yield erroneous temperatures, for example when the concentrations of major cations are not controlled primarily by feldspars and Mg phases such as chlorite and illite (assumptions on which most classical geothermometers rely, e.g., Giggenbach, 1988), or when fluids are affected by dilution and gas loss as they ascend to the surface from a deep reservoir, multicomponent geothermometry can offer a significant improvement over classical geothermometry. This is also true when geothermal fluids do not reach full chemical equilibrium with reservoir minerals, in which case the bracketing of temperatures using multiple minerals presents advantages over classical geothermometers, especially when minerals displaying both prograde and retrograde solubility behavior are selected for the analyses. This is not to say that multicomponent geothermometry cannot yield erroneous results. When applied carelessly, this method can also fail, such as when selecting inappropriate minerals or relying on Al and/or Mg concentrations that do not reflect reservoir conditions. In fact, in systems where fluids slowly cool (without significant mixing, boiling, or reaction with wall–rock minerals), the method is quite sensitive to the precipitation of small amounts of Al and Mg silicates minerals, because Al and Mg are typically present in trace amounts in solution, and their concentrations strongly affect the computation of saturation indices (e.g., Pang and Reed, 1998, Peiffer et al., 2014 and Wanner et al., 2014). In contrast, the Na–K and Na–K–Ca geothermometers are independent of the concentration of these trace elements and are expected to provide better results under cooling-only conditions (if feldspars primarily control the concentrations of these elements). Therefore, the strength of the geothermometry approach presented here is primarily for systems where deep chemical signatures have been masked by boiling, mixing, and/or reactions with minerals other than those on which classical geothermometers rely. When coupled with numerical optimization, the approach is quite powerful to reconstruct the composition of deep fluids and improve solute geothermometry, however the convergence of the method to an incorrect solution (local minima) can also yield erroneous results. Therefore, the method discussed in this study should never be applied blindly or taken as a new fail-proof solution to the challenging problem of solute chemical geothermometry. However, when used wisely, and in conjunction with classical geothermometry, the integrated multicomponent geothermometry approach developed here is expected to allow the assessment of target reservoir temperatures in a more reliable manner than previously achievable, which could translate to reduced costs of geothermal exploration and development.

ACKNOWLEDGMENTS

This work was supported by the U.S. Department of Energy, Geothermal Technologies Program, Energy Efficiency and Renewable Energy Office, Award No. DE-EE0002765. We thank

Patrick Dobson and Jennifer Lewicki for their valuable inputs after testing GeoT, Kevin Knauss for his leadership with the experimental component of this study, Joe Iovenitti (Alta Rock) for providing data on the Newberry project, and Christoph Wanner for compiling and testing GeoT on various platforms. We are also grateful to Stuart F. Simmons and William C. Evans for their constructive reviews which helped improve the original manuscript.

Appendix A: Water–rock Interaction Experiments

The water–rock interaction experiments used to test the multicomponent geothermometry method presented in this study are part of an on-going study aimed at describing the geochemical evolution of aqueous fluids in a natural geothermal system (Desert Peak, NV) as a function of time and water/rock ratio. These experiments were conducted in a Dickson-type rocking autoclave, consisting of a flexible gold bag, containing the reacting rock-powder and the fluid, sealed by commercially pure, passivated, Ti collar and Ti head. The reaction cell has a volume of ~240 ml and is hosted by a pressure vessel inside a large tube furnace that maintains the temperature constant through the entire duration of the experimental run. Pressure is held constant by a high-pressure syringe pump injecting distilled water into the pressure vessel, outside of the reaction cell. As aqueous samples are withdrawn, the gold bag collapses and an additional volume of distilled water is injected inside the vessel to counterbalance the reduction of volume due to the extraction of fluid.

Initial water to rock mass ratio varied between 5 and 215 and decreased typically by 35–40% at the end of the experiments, the total amount of fluid withdrawn from the gold cell during the experimental runs being taken into account. The aqueous fluid sampled from the reaction cell was filtered by an internal 10 μm Ti frit and then again using a 0.22 μm Millipore. Analyses of major cations and trace metals were carried out by ICP-OES and ICP-MS, respectively, whereas anions were measured by ion chromatography and total dissolved inorganic carbon (TDCI) was determined using a non dispersive infra-red (NDIR) CO_2 gas analyzer. The pH of the collected samples was measured at 25 °C using a traditional pH combination electrode.

REFERENCES

1. Arnorsson, S., Sigurdsson, S., Svavarsson, H., 1982. The chemistry of geothermal waters in Iceland. I. Calculation of aqueous speciation from 0 to 370 °C. Geochim. Cosmochim. Acta 46, 1513–1532.

2. Arnorsson, S., Gunnlaugsson, E., Svavarsson, H., 1983a. The chemistry of geothermal waters in Iceland. II. Mineral equilibria and independent variables controlling water compositions. Geochim. Cosmochim. Acta 47, 547–566.

3. Arnorsson, S., Gunnlaugsson, E., Svavarsson, H., 1983b. The chemistry of geothermal waters in Iceland. III. Chemical geothermometry in geothermal investigations. Geochim. Cosmochim. Acta 47, 567–577.

4. Blanc, Ph., Lassin, A., Piantone, P., Azaroual, M., Jacquemet, N., Fabbri, A., Gaucher, E.C., 2012. Thermoddem: a geochemical database focused on low temperature water/rock interactions and waste materials. Appl. Geochem. 27, 2107–2116.

5. Blanc, P., Lassin, A., Piantone, P., 2007. THERMODDEM, a database devoted to waste minerals. BRGM (Orléans, France) http://thermoddem.brgm.fr

6. Doherty, J., 2008. PEST – Model-Independent Parameter Estimation. Watermark Numerical Computing, Corinda 4075, Brisbane, Australia. http://www. sspa.com/pest/

7. Farrar, C.D., Sorey, M.L., Rojstaczer, S.A., Steinemann, A.C., Clark, M.D., 1989. Hydrologic and geochemical monitoring in Long Valley Caldera, Mono County, California, 1986. U.S. Geological Survey Water-Resources Investigations Report 89-4033.

8. Farrar, C.D., Sorey, M.L., Rojstaczer, S.A., Janik, C.J., Mariner, R.H., Winnett, T.L., Clark, M.D., 1985. Hydrologic and geochemical monitoring in Long Valley Caldera, Mono County, California, 1982–1984. U.S. Geological Survey Water-Resources Investigations Report 85-4183.

9. Finsterle, S., Zhang, Y., 2011. Solving iTOUGH2 simulation and optimization problems using the PEST protocol. Environ. Model. Softw. 26, 959–968, http://dx.doi.org/10.1016/j.envsoft.2011.02.008.

10. Fouillac, C., Michard, G., 1981. Sodium/lithium ratio in water applied to geothermometry of geothermal reservoirs. Geothermics 10, 55–70.

11. Fournier, R.O., Rowe, J.J., 1966. Estimation of underground temperatures from the silica content of water from hot springs and wet-steam wells. Am. J. Sci. 264, 685–697.

12. Fournier, R.O., Truesdell, A.H., 1973. An empirical Na–K–Ca geothermometer for natural waters. Geochim. Cosmochim. Acta 37, 1255–1275.

13. Fournier, R.O., Potter, R.W., 1982. A revised and expanded silica (quartz) geothermometer. Geotherm. Resour. Counc. Bullet. 11 (10), 3–12.

14. Fournier, R.O., 1977. Chemical geothermometers and mixing models for geothermal systems. Geothermics 5, 41–50.

15. Fournier, R.O., 1979. A revised equation for the Na/K geothermometer. Geotherm. Resour. Trans. 3, 221–224.

16. Giggenbach, W.F., 1988. Geothermal solute equilibria. Derivation of Na–K–Mg–Ca geoindicators. Geochim. Cosmochim. Acta 52, 2749–2765.

17. Holland, T.J.B., Powell, R., 1998. An internally consistent thermodynamic dataset for phases of petrological interest. J. Met. Geol. 16, 309–343.

18. Johnson, J.W., Oelkers, E., Helgeson, H.C., 1992. SUPCRT92: a software package for calculating the standard molal thermodynamic properties of minerals, gases, aqueous

species and reactions from 1 to 5000 bar and 0 to 1000 °C. Comput. Geosci. 1992 (18), 899–947.

19. Michard, G., Roekens, E., 1983. Modelling of the chemical composition of alkaline hot waters. Geothermics 12, 161–169.

20. Michard, G., Fouillac, C., Grimaud, D., Dennis, J., 1981. Une méthode globale d'estimation des températures des réservoirs alimentant les sources thermales. Exemple du Massif Central Francais. Geochim. Cosmochim. Acta 45, 1199–1207.

21. Palandri, J., Reed, M.H., 2001. Reconstruction of in situ composition of sedimentary formation waters. Geochim. Cosmochim. Acta 65, 1741–1767.

22. Pang, Z.-H., Reed, M.H., 1998. Theoretical chemical thermometry on geothermal waters: problems and methods. Geochim. Cosmochim. Acta 62, 1083–1091.

23. Peiffer, L., Wanner, C., Spycher, N., Sonnenthal, E., Kennedy, B.M., Iovenitti, J., 2014. Optimized multicomponent vs. classical geothermometry: insights from modeling studies at the Dixie Valley geothermal area. Geothermics 51, 154–169.

24. Powell, T., Cumming, W., 2010. Spreadsheets for geothermal water and gas geochemistry. In: Proceedings Thirty-Fifth Workshop on Geothermal Reservoir Engineering Stanford University, Stanford, California, February 1–3, SGP-TR- 188.

25. Reed, M.H., 1982. Calculation of multicomponent chemical equilibria and reaction processes in systems involving minerals, gases and an aqueous phase. Geochim. Cosmochim. Acta 46, 513–528.

26. Reed, M.H., 1997. Hydrothermal alteration and its relationship to ore fluid composition. In: Barnes, H.L. (Ed.), Geochemistry of Hydrothermal Ore Deposits. , 3rd ed. John Wiley & Sons, pp. 303–366.

27. Reed, M.H., 1998. Calculation of simultaneous chemical equilibria in aqueousmineral-gas systems and its application to modeling hydrothermal processes. In: Richards, J., Larson,

P. (Eds.), Techniques in Hydrothermal Ore Deposits Geology, Reviews in Economic Geology, vol. 10, pp. 109–124.

28. Reed, M.H., Palandri, J., 2006. SOLTHERM.H06, a database of equilibrium constants for minerals and aqueous species. Available from the authors, University of Oregon, Eugene, Oregon.

29. Reed, M.H., Spycher, N.F., 1984. Calculation of pH and mineral equilibria in hydrothermal waters with application to geothermometry and studies of boiling and dilution. Geochim. Cosmochim. Acta 48, 1479–1492.

30. Sandia National Laboratories, 2007. Qualification of thermodynamic data for geochemical modeling of mineral–water interactions in dilute systems (data0.ymp.R5). Report ANL-WIS-GS-000003 REV 01. Sandia National Laboratories, Las Vegas, Nevada, ACC: DOC.20070619.0007.

31. Shock, E., Sassini, D., Willis, M., Sverjensky, D., 1997. Inorganic species in geologic fluids: correlations among standard molal thermodynamic properties of aqueous ions and hydroxide complexes. Geochim. Cosmochim. Acta 61, 907–950.

32. Sonnenthal, E., Spycher, N., Callahan, O., Cladouhos, T., Petty, S., 2012. A thermal–hydrological–chemical model for the Enhanced Geothermal System Demonstration Project at Newberry Volcano, Oregon. In: Proceedings ThirtySeventh Workshop on Geothermal Reservoir Engineering Stanford University, Stanford, California, SGP-TR-194.

33. Spycher, N., Sonnenthal, E., Kennedy, B.M., 2011. Integrating multicomponent chemical geothermometry with parameter estimation computations for geothermal exploration. Geotherm. Resour. Counc. Trans. 35, 663–666.

34. Tempel, R.N., Sturmer, D.M., Schilling, J., 2011. Geochemical modeling of the nearsurface hydrothermal system beneath the southern moat of Long Valley Caldera, California. Geothermics 40 (2011), 91–101.

35. Verma, S.P., Pandarinath, K., Santoyo, E., 2008. SolGeo: a new computer program for solute geothermometers and its application to Mexican geothermal fields. Geothermics 37, 597–621.

36. Wanner, C., Peiffer, L., Sonnenthal, E., Spycher, N., Iovenitti, J., Kennedy, B.M., 2013. Assessing thermo-hydrodynamic-chemical processes at the Dixie Valley geothermal area: a reactive transport modeling approach. In: Proceedings 38th Workshop on Geothermal Reservoir Engineering, Stanford Univ. Report SGP-TR-198.

37. Wanner, C., Peiffer, L., Sonnenthal, E., Spycher, N., Iovenitti, J., Kennedy, B.M., 2014. Reactive transport modeling of the Dixie Valley geothermal area: insights on flow and geothermometry. Geothermics 51, 130–141.

38. Wolery, T.J., 1979. Calculation of chemical equilibrium between aqueous solution and minerals; the EQ3/6 software package. UCRL-52658. Lawrence Livermore Laboratory.

39. Xu, T., Spycher, N., Sonnenthal, E., Zhang, G., Zheng, L., Pruess, K., 2011. TOUGHREACT Version 2.0: a simulator for subsurface reactive transport under non-isothermal multiphase flow conditions. Comput. Geosci. 37, 763–774.

40. Xu, T., Sonnenthal, E., Spycher, N., Pruess, K., 2006. TOUGHREACT: a simulation program for non-isothermal multiphase reactive geochemical transport in variably saturated geologic media: applications to geothermal injectivity and CO2 geological sequestration. Comput. Geosci. 32, 145–156.

Chapter 6

Effects of Gas Types and Models on Optimized Gas Fuelling Station Reservoir's Pressure

M. Farzaneh-Gord, M. Deymi-Dashtebayaz,
and H. R. Rahbari

The Faculty of Mechanical Engineering, Shahrood University of Technology, Zip Code 3619995161, Shahrood, Iran.n

ABSTRACT

There are similar algorithms and infrastructure for storing gas fuels at CNG (Compressed Natural Gas) and CHG (Compressed Hydrogen Gas) fuelling stations. In these stations, the fuels are usually stored in the cascade storage system to utilize the stations more efficiently. The cascade storage system generally divides into three reservoirs, commonly termed low, medium and high-pressure reservoirs. The pressures within these reservoirs have huge effects on performance of the stations. In the current study, based on the laws of thermodynamics, conservation of mass and real/ideal gas assumptions, a theoretical analysis has been constructed to study the effects of gas types and models on performance of the stations. It is intended to determine the optimized reservoir pressures for these stations. The results reveal that the optimized pressure differs between the gas types. For ideal and real gas models in both stations (CNG and CHG), the optimized non-dimensional low pressure-reservoir pressure is found to be 0.22. The optimized non-dimensional medium-pressure reservoir pressure is the same for the stations, and equal to 0.58.

INTRODUCTION

There are a lot of natural gas vehicles (NGV) in use, and the number is expanding every year. A number of national and international standards have been issued to ensure safe and efficient use of NGVs. Due to the relatively small number of hydrogen fuelled vehicles (HGV) currently on the roadway; there are limited regulations for these vehicles. The HGVs are currently using similar standards, regulations, infrastructures and fuelling stations to NGVs.

In these stations, the vehicles usually receive fuels from high pressure reservoirs during filling. The first problem with these fuel stations is the refuelling time. The NGV and HGV industries have made excellent advancements to provide a system to refuel a NGV or a HGV in a time comparable to that of a gasoline station. This fill time can be referred to as a fast fill or rapid charge.

The on-board storage capacity of natural gas and hydrogen vehicles is the other problem for the widespread marketing of these alternate fuelled vehicles. The on-board storage cylinders undergo a rise in storage gas cylinder temperature in the range of 40 K or more for CNG (Kountz, 1994) and 70 K or more for CHG (Dicken et al., 2007) during fast filling. For both fuels (CNG and

CHG), this temperature rise reduces the density of the gas in the cylinder, resulting in an under-filled cylinder relative to its rated specification. If this temperature rise is not compensated for in the fuelling station dispenser, by transiently over-pressurizing the tank, the vehicle user will experience a reduced driving range. Although NGV and HGV on-board cylinder volumes play the main role in the on-board storage capacity, but the fuelling station reservoir's pressure also has big effects on the amount of the filled mass of the on-board cylinder.

Natural gas from the distribution pipeline is compressed using a large multi-stage compressor into a "cascade" storage system. On the other hand, hydrogen must also be compressed by a compressing system and stored in a "cascade" storage system (Baur et al., 2004). The input work required by the compressor is the final problem with the fuelling stations, as the required work for compressing and storing the natural gas and hydrogen are then partially wasted through the filling process.

In order to make the utilization of the natural gas and hydrogen fuelling stations more efficient, these fuels are usually stored in a cascade storage system. The cascade storage system is commonly divided into three reservoirs, generally termed low (LPR), medium (MPR) and high-pressure reservoirs (HPR). The pressure within the reservoirs has big effects on 3 problems associated with the filling process and fuelling station as below:

- Filling time;
- The charged mass of the on-board cylinder after refuelling;
- The compressor input work.

Considering the above three problems with a compressed gas fuelling station, one can conclude that, by reducing filling time, reducing compressor input work and (or) increasing the filled mass of the on-board cylinder, the performance of the compressed gas fuelling stations would be improved.

There is a strong possibility of using the current natural gas infrastructure as a starting point for hydrogen vehicle infrastructure. In this research, the effects of important parameters on the performance of CNG and CHG stations are studied. The principal aim is to show that the optimum algorithm for these stations is slightly different.

To understand the fast filling process and study the effect of the pressure and temperature within low, medium and high-pressure reservoirs on the performance of CNG and CHG fuelling stations, a theoretical analysis has been developed in this study. The fast fill process was assumed to be a quasi-static process and the natural gas presumed to be purely methane (as an ideal and real gas). The hydrogen is also treated as an ideal and real gas. A second law analysis has been employed to calculate the amount of entropy generations during the filling process. It is well known that a lower entropy generation is associated with less work required work by the compressor.

There have been limited researches in the field of filling process modelling in the literature. Kountz et al. (1997) were the first to model the fast filling process of a natural gas storage cylinder based on the first law of thermodynamics. They developed a computer program to model the fast filling process for a single reservoir. They have also developed a natural gas dispenser control algorithm that insures complete filling of NGV cylinders under a fast-fill scenario (Kountz et al., 1998; Kountz et al., 1998; Kountz et al., 1998). Research has also been carried out to model fast-filling of hydrogen-based fuelling infrastructure. Fast filling of a hydrogen cylinder using a number of experiments (Liss et al., 2002; Liss et al., 2003; Newhouse et al., 1999) resulted in a high temperature increase in the cylinder during the process. In another research, a control method optimized for a high utilization ratio and fast filling speed in hydrogen refuelling stations was reported (Zheng et al., 2010). The results of this research show that an optimized control method can significantly improve the utilization ratio and allows refuelling in an acceptable time.

A few experimental studies were also carried out to study fast filling of natural gas on-board cylinders (Thomaset al., 2002; Shiply, 2002). Shiply (2002) concluded that ambient temperature variation could affect the fast-fill process. He also concluded that the test cylinder is under-filled every time it is rapidly recharged. For the hydrogen fast fuelling process, there have been experiments on thermal characteristics during the hydrogen filling process of type IV cylinders (Chan Kim et al., 2010). In that study, a computational fluid dynamics (CFD) analysis was also employed to simulate the conditions of the experiments. The CFD results show reasonable agreement with the experiments. The discrepancy between the CFD and experimental values decreases as the initial gas pressure increases.

The authors of the current study have also modelled the CNG fast filling process (Farzaneh-Gord et al. 2007; Farzaneh-Gord, 2008, Farzaneh-Gord et al. 2013, Deymi-Dashtebayaz et al., 2013). They developed a computer programme based on the Peng-Robinson state equation and methane properties for a single reservoir fuelling station. They investigated the effects of ambient temperature and initial cylinder pressure on the final cylinder conditions. In another study, they (Farzaneh-Gord et al. 2008) presented thermodynamics analysis of the cascade reservoir filling process of natural gas vehicle cylinders. The results indicated that ambient temperature has a big effect on the filling process and final NGV cylinder conditions. Farzaneh-Gord et al. (2011) have employed a theoretical analysis to study the effects of buffer and cascade storage systems on performance of a CNG fuelling station. It was found that the time (filling time) required for bringing up the NGV on-board cylinder to its final pressure in the buffer storage system is about 66% less than the cascade storage system. The charged mass for the cascade system is about 80% of the buffer system, which gives an advantage to the buffer system over the cascade system. The biggest advantage of the cascade system over the buffer system is 50% less entropy generation for this configuration, which probably causes much lower required compressor input work for this configuration compared to the buffer system. Farzaneh-Gord et al. (2012a, 2012b) have also studied effects of natural gas compositions on fast filling process for buffer and cascade storage banks. In these researches, the conditions of storage bank were considered constant. Farzaneh-Gord et al. (2012c) have examined two storage systems for compressed hydrogen stations. The important parameters such as filling time, filled mass and compressor input work have been examined in detail. The results show that the filling time of the buffer storage system is much less than that of the cascade storage system. However, the filled mass related to the buffer system for the same conditions is approximately equal to that of the cascade system. Furthermore, the buffer system is accompanied by much higher entropy generation as compared to the cascade storage system, which is directly reflected in the amount of compressor input work required.

As mentioned previously, the second law has been employed in this study to calculate the entropy generation theoretically. Entropy generation is associated with thermodynamic irreversibilities, which are common in all types of thermal systems. There have been numerous investigations in the field of entropy generation. A researcher has concentrated on the different mechanisms responsible for entropy generation in applied thermal engineering (Bejan, 1982; 1996). Generation of entropy dissipates work into heat. Therefore, it makes good engineering sense to focus on irreversibilities of heat transfer and fluid flow processes and try to understand the function of the related entropy generation mechanisms (Bejan, 1972). Since then, a lot of investigations have been carried out to compute the entropy generation and irreversibility profiles for different geometric configurations, flow situations, and thermal boundary conditions (Sordi et al., 2009; Diaz et al., 2007). Here, entropy generation has been employed as a main tool to determine the amount of work destruction during filling.

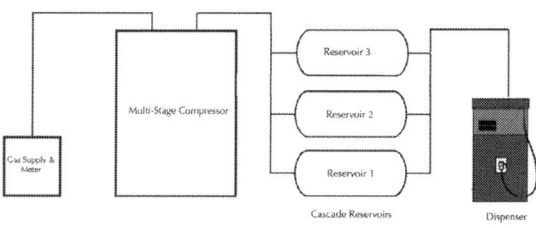

Figure 1: A schematic diagram of a NGV filling station.

CNG AND CHG FILLING STATIONS

Figure 1 shows a typical CNG filling station. Gas from the distribution pipeline, usually "low" pressure at <4bar (0.4 MPa) or possibly "medium" pressure (1.6 MPa), is compressed using a large multi-stage compressor into a "cascade" storage system. The pressure within the storage system is kept at a higher value than in the vehicle's on-board cylinder so that gas transfers to the vehicle under differential pressure. Commonly, the cascade storage will operate in the range of 20.5 MPa to 25 MPa, while the vehicle's maximum on-board cylinder pressure is 20 MPa. For more efficient use of the compressor and the storage system, the CNG stations commonly operate using a three-stage "cascade" storage system.

In a CHG fuelling station, there is a similar algorithm as in a CNG fuelling station. The hydrogen can be supplied using various methods (Sordi et al., 2009). The common method for supplying hydrogen is by using an electrolyser (Baur et al., 2004). In this method, hydrogen can be produced through the electrolysis of water. The hydrogen is then compressed to a high pressure similar to a CNG station. In CHG stations, the cascade storage operates in the range of 35-44 MPa (or 70-80 MPa), while the vehicle's maximum on-board cylinder pressure is 35 MPa (or 70 MPa) (Zheng et al., 2010).

The cascade storage system is usually divided into three reservoirs, commonly termed low, medium and high-pressure reservoirs (Thomas et al., 2002). During fast filling, the on-board cylinder is first connected to the low-pressure reservoir. As the pressure in the reservoir falls and that in the onboard cylinder rises, the flow of gas decreases. When the flow rate has declined to a pre-set level, the system switches to the medium pressure reservoir, then finally to the high-pressure reservoir to complete the fill. It is expected that the cascade system results in a more complete "fill" than if the whole storage were maintained at one pressure and utilises the compressor and storage with maximum efficiency. In addition, when the compressor is automatically turned on to refill the reservoirs it fills the high pressure reservoir first, and then switches to the medium and the low reservoirs. This ensures that the high pressure reservoir (employed to complete the fill) is kept at maximum pressure all the time, ensuring that vehicles are always supplied with the maximum amount of gas available. Correct specification of the compressor capacity and the volume of cascade storage is necessary to ensure that the CNG and CHG stations can cope with the type (passenger cars, buses or trucks) and frequency (peak periods) of vehicles using the facility.

Compressed Natural Gas and Hydrogen Cylinders

The natural gas and hydrogen cylinders have various design types based on the construction materials used. Design types include Type 1, which are all-metal, Type 2, which have a metal liner and a hoop-wrapped composite reinforcement, Type 3, which have a metal liner and a full-wrapped composite reinforcement, and Type 4, which have a non-metallic liner and a full-wrapped composite reinforcement. Metal containers and liners are typically steel or aluminium. Composite reinforcements are typically carbon or glass fibers in an epoxy resin matrix. These cylinders are designed for a specified nominal service pressure at a specified temperature, essentially a specified density (kg/m^3) of fuel. This will result in a given mass of natural gas or hydrogen stored in the fuel container (cylinders). The actual pressure in the fuel container will vary from the nominal service pressure as the temperature of the fuel in the cylinder varies. Under-filling of the on-board cylinders can occur at fuelling stations during fast-fill charging operations at ambient temperatures greater than 30 °C. The resulting reduced driving range of the vehicle is a serious obstacle that the gas industry is striving to overcome, without resorting to unnecessarily high fuelling station pressures, or by applying extensive over-pressurization of the cylinder during the filling operation. Undercharged cylinders are a result of the elevated temperature that occurs in the CHG and CNG storage cylinder.

Chemical Compositions of Natural Gas

Natural gas composition (mixture) varies with location, climate and other factors. The gas is refined before flowing into the pipe lines. Methane is the component that is a very high percentage of natural gas. As a result and for the sake of simplicity, it is assumed that methane is the only substance in the natural gas.

Cascade Reservoir Parameters

For a cascade storage system, there are a few important parameters that affect the filling. These parameters are introduced in this section.

Figure 2 shows a schematic diagram of a cascade reservoir system. Thermodynamic properties in cascade reservoirs play important roles in the filling process. Two main properties are pressure and temperature. As shown in Figure 2, each reservoir has its own temperature (TR) and pressure (PR). These are assumed to be unchanged while the pressure and temperature in the on-board NGV and HGV cylinders varies during filling.

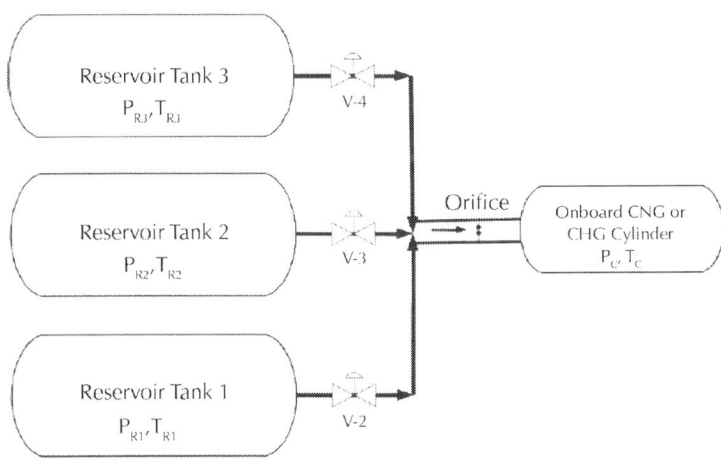

Figure 2: A schematic diagram of cascade reservoirs.

To maintain the final pressure within the onboard cylinder at its rated value, the pressure within the HPR assumed to be constant (20.5 MPa for CNG and 37 MPa for CHG) throughout this study. The effects of medium and low reservoir pressures on performance of the fuelling station have been studied by introducing the two dimensionless parameters. The ratio of the medium and low pressure reservoirs to the high-pressure one, defined as below, are these two parameters:

$$NP1 = P_{R1} / P_{R3} \qquad NP2 = P_{R2} / P_{R3} \tag{1}$$

The final dimensionless number is the "fill ratio". The cylinder "fill ratio" is defined as the mass of charged gas after refuelling divided by the mass which the cylinder could hold at the rating condition (300 K, 20 MPa for CNG and 300 K, 35 MPa for CHG). This parameter is directly related to the driving range of the NGV and HGV, defined as:

$$FR = \frac{m_c \text{ (at end of filling)}}{\rho(300K, 20MPa \text{ for CNG and } 300K, 35MPa \text{ for CHG})V_c} \tag{2}$$

It should also be noted that the cylinder volumes (V_c) for NGV and HGV were considered to be 67 and 150 liters (Zheng et al., 2010), respectively, throughout this study.

THERMODYNAMIC ANALYSIS

First Law Analysis

In this study, in order to model the fast filling process and develop a mathematical method, the onboard cylinder was considered to be an open thermodynamic system which undergoes a quasisteady process.

To develop a theoretical analysis, continuity and the first law of thermodynamics have been applied to the cylinder to find 2 thermodynamic properties. Considering the onboard cylinder as a control volume based onFigure 3 and knowing that it has only 1 inlet, the continuity (conservation of mass) equation may be written as follows:

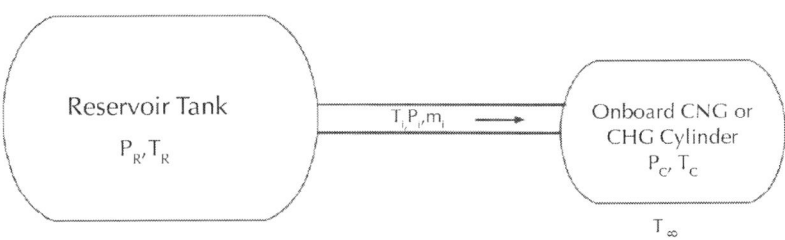

Figure 3: A schematic diagram of the thermodynamic system.

$$\frac{dm_C}{dt} = \dot{m}_i$$

(3)

In Equation (3), $\dot{m}_i$ is the inlet mass flow rate and can be calculated by considering an isentropic expansion through an orifice. Applying gas dynamics laws:

$$\dot{m}_i = C_d \rho_R A_{orifice} \left(\frac{P_C}{P_R}\right)^{\frac{1}{\gamma}}$$

$$\left\{\left(\frac{2\gamma}{\gamma-1}\right)\left(\frac{P_R}{\rho_R}\right)\left[1-\left(\frac{P_C}{P_R}\right)^{\frac{\gamma-1}{\gamma}}\right]\right\}^{\frac{1}{2}} \quad if \; \frac{P_C}{P_R} \leq \left(\frac{2}{\gamma+1}\right)^{\frac{\gamma}{\gamma-1}}$$

(4)

$$\dot{m}_i = C_d \sqrt{\gamma P_R \rho_R} \, A_{orifice} \left(\frac{2}{\gamma+1}\right)^{\frac{\gamma+1}{2(\gamma-1)}}$$

$$if \; \frac{P_C}{P_R} > \left(\frac{2}{\gamma+1}\right)^{\frac{\gamma}{\gamma-1}}$$

(5)

In Equations (4)-(5), C_d is the discharge coefficient of the orifice. The first law of thermodynamics in general form can be written as:

$$\dot{Q}_{cv} + \sum \dot{m}_i (h_i + V_i^2 / 2 + gz_i)$$
$$= \sum \dot{m}_e (h_e + V_e^2 / 2 + gz_e)$$
$$+ d / dt[m(u + V^2 / 2 + gz)]_{cv} + \dot{W}_{cv} \tag{6}$$

The work term is zero in the filling process and the change in potential and kinetic energy can be neglected. The equation can then be simplified as below:

$$\frac{dU_C}{dt} = \dot{Q} + \dot{m}_i (\frac{V_i^2}{2} + h_i) \tag{7}$$

Since $h_R = V_i^2 / 2 + h_i$, the above equation can be further simplified as:

$$\frac{dU_C}{dt} = \dot{Q} + \dot{m}_i h_R \tag{8}$$

The heat lost from the onboard cylinder to the environment can be calculated as

$$\dot{Q} = -U_{HC} A_C (T_C - T_\infty) \tag{9}$$

Combining (3), (8) and (9), one can write the above equation as:

$$\frac{d(m_C u_C)}{dt} = -U_{HC} A_C (T_C - T_\infty) + \frac{dm_C}{dt} h_R \tag{10}$$

Or in the following form:

$$\frac{d(m_C u_C)}{dt} - \frac{d(m_C h_R)}{dt} = -U_{HC} A_C (T_C - T_\infty) \tag{11}$$

The above equation can be rearranged to the following form:

$$d(m_C u_C - m_C h_R) = -U_{HC} A_C (T_C - T_\infty) dt \tag{12}$$

The above equation can be integrated from the "start" of filling up to the "current" time as:

$$\int_s^c d(m_C u_C - m_C h_R) = -\int_0^t U_{HC} A_C (T_C - T_\infty) dt$$

(13)

The integration of the above equation for a single reservoir fuelling station results in:

$$m_C(u_C - h_R) - m_{Cs}(u_{Cs} - h_R) = -U_{HC} A_C \Delta T_{av} t$$

(14)

Where, m_c, m_{cs} are the mass of charged gas at the "current" and "start" of the filling process, ΔT_{av} is the average temperature difference between the cylinder and environment, which is defined as:

$$\Delta T_{av} = \frac{1}{t} \int_0^t (T_C - T_\infty) dt$$

(15)

The first law of thermodynamics for the onboard cylinder can finally be written as:

$$u_C = h_R - \frac{U_{HC} A_C \Delta T_{av} t}{m_C} + \frac{m_{Cs}}{m_C}(u_{Cs} - h_R)$$

(16)

The Equation (3), (4) and (16) can be employed to calculate the two thermodynamic properties of the in-cylinder natural gas and hydrogen at any time. By knowing two thermodynamics properties (here the specific internal energy and specific volume), the other in-cylinder properties could be found.

Adiabatic System

For an adiabatic system, Equation (16) can be further simplified as:

$$u_C = h_R + \frac{m_{Cs}}{m_C}(u_{Cs} - h_R)$$

(17)

And if, $m_{cs}=0$, the following relation is valid at any time:

$$u_C = h_R$$

(18)

Ideal Gas Model for an Adiabatic System

For the case of ideal gas behaviour, the governing equation can be substantially simplified. Considering the following ideal gas assumptions:

$$u = c_v T, h = c_p T, m = \frac{PV}{RT}$$

(19)

And knowing that the volume of the cylinder, specific heats, and reservoir temperature are constant, then Equation (8) can be simplified as follows:

$$\frac{d(mu)_{cv}}{dt} = \dot{m}_i h_R \Rightarrow \frac{d\left(\frac{PV}{RT} c_v T\right)_{cv}}{dt} = \dot{m}_i c_p T_R$$

$$\Rightarrow \frac{V_{cv} c_v}{R} \frac{d(P_{cv})}{dt} = \dot{m}_i c_p T_R$$

(20)

By replacing the inlet mass flow rate from Equations (4) and (5), the following simple equation can be obtained:

$$d(P_C)/dt = m_i(\gamma R / V_{cv})T_r = \begin{cases} (\gamma R / V_{cv})T_R C_d P_R A_{onfice}\left(\frac{P_C}{P_R}\right)^{\frac{1}{\gamma}}\left[\left(\frac{2\gamma}{\gamma-1}\right)\left(\frac{P_R}{\rho_R}\right)\left[1-\left(\frac{P_C}{P_R}\right)^{\frac{\gamma-1}{\gamma}}\right]\right]^{\frac{1}{2}} & \text{if } \frac{P_C}{P_R} \le \left(\frac{2}{\gamma+1}\right)^{\frac{\gamma}{\gamma-1}} \\ (\gamma R / V_{cv})T_R C_d \sqrt{\gamma P_R \rho_R} A_{onfice}\left(\frac{2}{\gamma+1}\right)^{\frac{\gamma+1}{2(\gamma-1)}} & \text{if } \frac{P_C}{P_R} > \left(\frac{2}{\gamma+1}\right)^{\frac{\gamma}{\gamma-1}} \end{cases}$$

(21)

The Second Law Analysis

The second law of thermodynamics and the flow processes occurring in the "cascade" storage system of the gas filling station adopted in this study makes it possible to evaluate the entropy generation rates, $\dot{S}_{gen}$ for the characteristic nodes of the system.

The second law of thermodynamics for the filling process of the on-board cylinder can be expressed as:

$$\dot{S}_{gen} = dS_C / dt - \dot{Q} / T_\infty - \dot{m}_i s_i \ge 0$$

(22)

Here, all irreversibility is assumed to occur at the inlet to in-cylinder position. There is an isentropic expansion from the reservoir to the inlet position, which means. $s_i = s_R$ considering this assumption and combining Equation (3), (9) and (22), the following equation can be obtained:

$$\dot{S}_{gen} = \frac{d(m_C s_C)}{dt} - \frac{dm_C}{dt} s_R$$

$$+ U_{Hc} A_C (T_C - T_\infty) / T_\infty \tag{23}$$

In the following form:

$$\dot{S}_{gen} dt = d(m_C s_C - m_C s_R)$$

$$+ U_{HC} A_C (T_C - T_\infty) / T_\infty dt \tag{24}$$

The above equation can be integrated from "start" of filling to the "current" time as below:

$$S_{gen} = \int_s^c d(m_C s_C - m_C s_R)$$

$$+ \int_s^c \frac{U_{HC} A_C (T_C - T_\infty)}{T_\infty} dt \tag{25}$$

For a fuelling station with a single reservoir in which s_R remains constant throughout the filling process, the integration of the above equation results in a simple equation:

$$S_{gen} = m_C (s_C - s_R) - m_{Cs} (s_{Cs} - s_R)$$

$$+ \frac{U_{HC} A_C (T_{av} - T_\infty)}{T_\infty} \tag{26}$$

Adiabatic System

Equation (26) can be further simplified for an adiabatic system as:

$$S_{gen} = m_C (s_C - s_R) - m_{Cs} (s_{Cs} - s_R) \tag{27}$$

And if the cylinder is empty at the start of the filling process ($m_{cs} = 0$), the following relation can be obtained:

$$S_{gen,max} = m_C(s_C - s_R)$$

(28)

Ideal Gas Model for an Adiabatic System

For the case of ideal gas behaviour, the second law is much simplified. Considering the following ideal gas assumptions:

$$s_C - s_R = c_p \ln\frac{T_C}{T_R} - R\ln\frac{P_C}{P_R},$$

$$m_c = \frac{P_C V_C}{RT_C}$$

(29)

And knowing that the volume of the cylinder, V_c specific heat, c_p and reservoir temperature are constant, then Equation (27) can be simplified as follows:

$$S_{gen} = m_C(c_p \ln\frac{T_C}{T_R} - R\ln\frac{P_C}{P_R})$$

$$- m_{Cs}(c_p \ln\frac{T_{Cs}}{T_R} - R\ln\frac{P_{Cs}}{P_R})$$

(30)

For the case of $m_{cs}=0$ and assuming that the pressure within the on-board cylinder reaches its reservoir pressure, $(p_C \approx p_R)$ then Equation (30) gives the maximum entropy generation for a fuelling station as follows:

$$S_{gen,max} = S_{gen} = m_C c_p \ln\frac{T_c}{T_R}$$

(31)

Considering Equations (18) and (19), the above equation can be further simplified as follows:

$$S_{gen.max} = S_{gen} = \frac{c_v P_R V_c}{R T_R} \ln \gamma$$

(32)

It should be noted that Equations (29) to (32) are only valid for a single reservoir fuelling station. Calculating the entropy generation for a fuelling station with a cascade reservoir system demands more effort. Here the non-dimensional entropy generation is introduced to compare the results for various configurations as follows:

$$NS = S_{gen} / S_{gen.max}$$

(33)

It is worth mentioning that expresses the irreversibility in the system. The minimization of means reducing the part of input work which dissipates into heat in the system. As all work required by the system is provided by the station compressor, one can conclude that the minimum of indicates the least input work by the compressor.

SIMULATION METHODOLOGY

The set of equations presented in the previous section constitutes the model for the filling. The solution of these equations is, however, not straight-forward. The simulation proceeds in the following steps:

- Initial reservoir and on-board cylinder thermodynamic properties are known.
- Equation (4) (or (5)) is employed to calculate the mass flow rate.
- By assuming a small time increment, Equation (2) is solved numerically to calculate the in-cylinder mass (and specific volume) at the new time step.
- The energy equation is solved numerically to calculate the specific internal energy at the new time step.
- By knowing the specific volume and specific internal energy and utilizing the properties table, other thermodynamic properties will be found by trial and error.

- If the in-cylinder pressure has reached its target value, the simulation is stopped, otherwise go to step 2.

RESULTS AND DISCUSSION

In this study, the NGV and hydrogen cylinders have been considered to be adiabatic. Consequently, the orifice diameter will affect the final state in the cylinder. The orifice diameter will affect the filling time and inlet mass flow rate. The orifice diameter and the cylinder volume for NGV and HGV were considered to be 1 mm and 67 and 150 liters (Zheng et al., 2010), respectively. Also in this study ambient temperature was fixed at 300 K. The results are presented here for the commonly used cascade group of three, as shown in Figure 2. Figure 4 shows dynamic pressure profiles for CNG and CHG cylinders during filling for the real and ideal gas models. The LPR, MPR, HPR pressures are assumed to be respectively 5.5, 10.5 and 20.5 MPa for CNG and 11, 21 and 37 MPa for CHG. Discontinuity in the pressure profile is due to switching to another reservoir tank. As expected, the final in-cylinder pressure for CNG and CHG cylinders for both gas models is constant, respectively 20 and 35 MPa.

Figure 5 shows dynamic in-cylinder mass profiles for CNG and CHG cylinders during filling. In this figure, the thermodynamic conditions of reservoirs are similar to Figure 4. Discontinuity in the profiles is due to switching to another reservoir tank. Note from the figure that the final CHG in-cylinder mass for the ideal gas model is much higher than for the real gas model, whereas the final CNG in-cylinder mass for the real gas model is much higher than for the ideal gas model. The reason is due to variations in the final in-cylinder temperature. As the final in-cylinder pressure reaches its targeted value, the in-cylinder mass is higher for the cylinder in which the final temperature is lower. Yang (2009) carried out a theoretical analysis and calculated the final in-cylinder temperature to initial temperature ratio for a single reservoir system for CHG. The ratio was found to be 1.48.

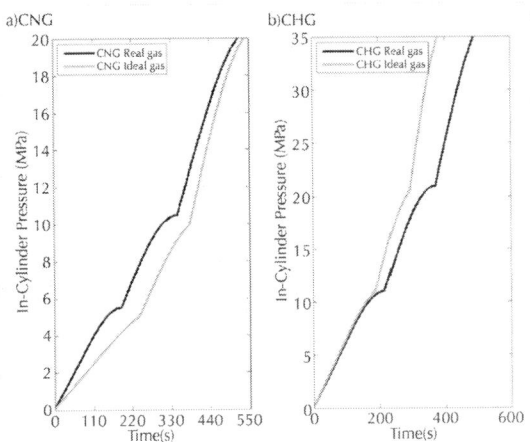

Figure 4: CNG and CHG dynamic in-cylinder pressures for the real and ideal gas models.

In the present study, the ratio was found to be 1.42 and 1.45 for the ideal and real CHG gas models, respectively.

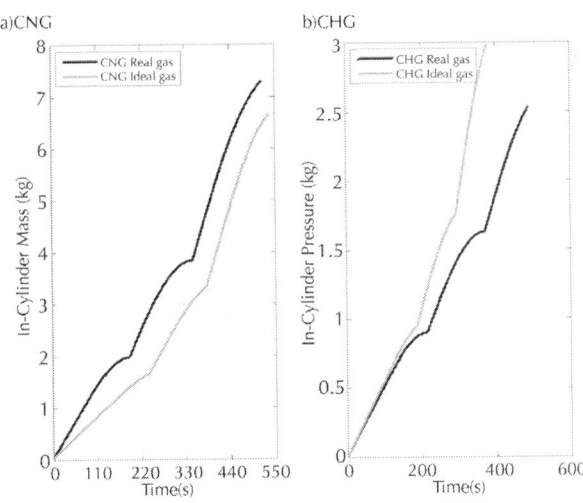

Figure 5: CNG and CHG dynamic in-cylinder masses for the real and ideal gas models.

Figure 6 shows dynamic in-cylinder temperature profiles for CNG and CHG during filling for the real and ideal gas models. Note from this figure that, for CNG, there is no dip in the temperature profile at early filling times; however, two dips in the temperature profile occur when the supply tank changes. This behaviour can indicate that the Joule-Thompson cooling effect is not high enough in the early stage of the filling process and is not able to overcome the conversion of reservoir enthalpy energy into cylinder internal energy. When the supply system switches to the higher pressure reservoir tank, the higher Joule-Thompson cooling effect, with the help of the low temperature inlet gas, causes small dips in the temperature profile. Because the Joule-Thompson coefficient for CHG is positive throughout the filling, the temperature will rise as the pressure drops. This causes the final in-cylinder temperature for real CHG to be higher than that of ideal CHG.

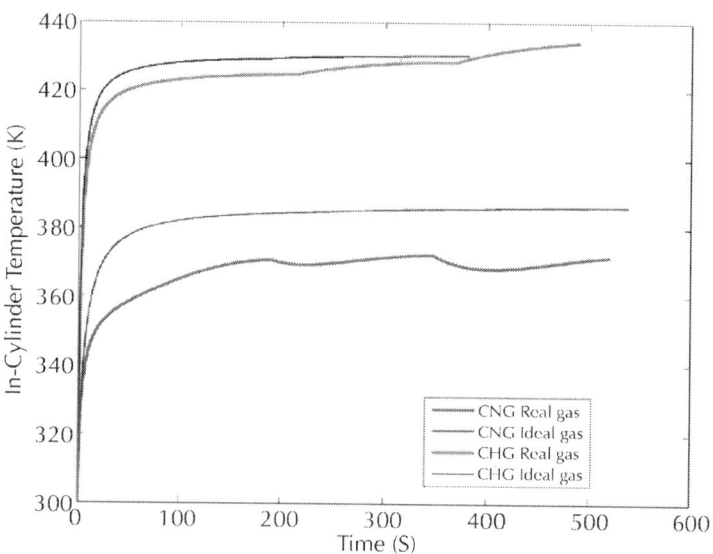

Figure 6: CNG and CHG dynamic in-cylinder temperatures for the real and ideal gas models.

Note from Figure 6 that the rise in temperature only occurs when the CNG and CHG cylinders are connected to the lowest pressure reservoir tank and the temperature profile is nearly monotonic otherwise. Note that the discontinuity in the temperature profile in Figure 6 is due to switching to another reservoir tank.

As expected for the ideal gas model, there is no dip in the temperature profile due to the fact that the Joule-Thompson cooling effect is not present. Comparing the real and ideal models in Figure 6, it can be realized that the temperature profiles are highly different and the temperature rise for CHG is much higher than for CNG cylinders. So it can be concluded that the thermodynamic properties of the gas have a big effect on the temperature profile.

The results presented here are only valid for adiabatic on-board cylinders. In real conditions, the on-board cylinders are not adiabatic. This makes the final in-cylinder temperature lower than the values reported in the present study due to the heat lost during the filling. The actual final in-cylinder temperature depends on the charging time, but for safety considerations it should be lower than 85 °C. To avoid this problem, the charging time is controlled by the station dispenser algorithm through regulating the inlet mass flow rate.

Figure 7 shows how the fill ratio varies with initial temperature (in the cylinders), which could represent the effect of ambient temperature. It can be seen that as the initial temperature increases the fill ratio decreases. This means that the driving range of CNG and CHG vehicles will decrease if filling is carried out in hot weather compared to the colder conditions. The ambient temperature has opposite effects on the final in-cylinder temperature. The final in-cylinder temperature increases as ambient temperature increases. Note again from Figure 7 that the fill ratio is higher for the real gas model in the CNG cylinder. For CHG, the fill ratio is higher for the ideal gas model.

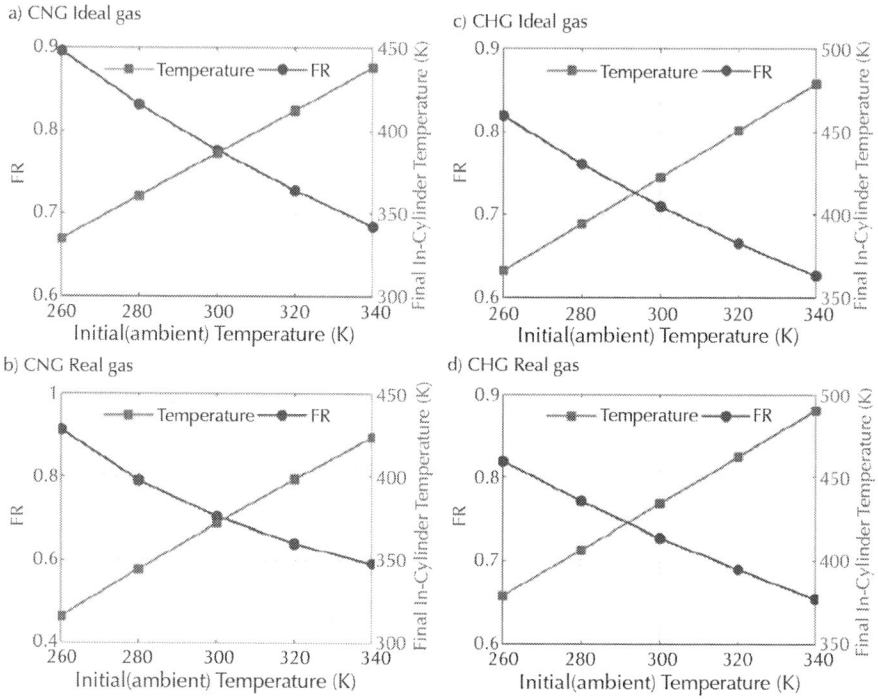

Figure 7: Effect of initial (ambient) temperature on fill ratio and final in-cylinder temperature for the ideal and real gas models.

Figure 8 shows the effect of initial cylinder and reservoir tank temperature on the filling time for the real and ideal gas models in CNG and CHG stations. As can be seen, as initial temperature increases, filling time decreases. Note from the figure that, for CNG, the filling time for the real gas model is less than for the ideal gas model. This is mainly due to less mass filled into the cylinder (see FR in Figure 7). For CHG, the FR is higher for the real gas model, so it is expected that the filling time will be higher for the ideal gas model. This can be seen in the figure.

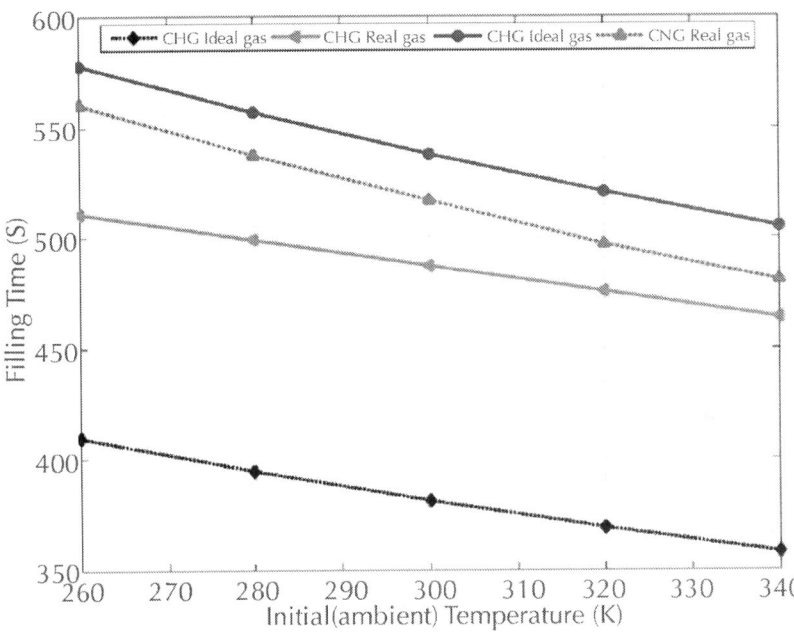

Figure 8: Effect of initial (ambient) temperature on filling time (NP2 = 0.53) for the ideal and real gas models in the two stations (CNG and CHG).

Figure 9 shows the effects of LPR pressure on filling time when the MPR pressure is kept constant (NP2 = 0.53) for the real and ideal gas models in CNG and CHG stations. As the LPR pressure (NP1) increases, the filling time decreases except for the ideal CNG case. For the ideal gas model in a CNG station, the maximum filling time occurs at NP1 ≈ 0.17. As the LPR pressure (NP1) increases, the filling times approach each other for all cases.

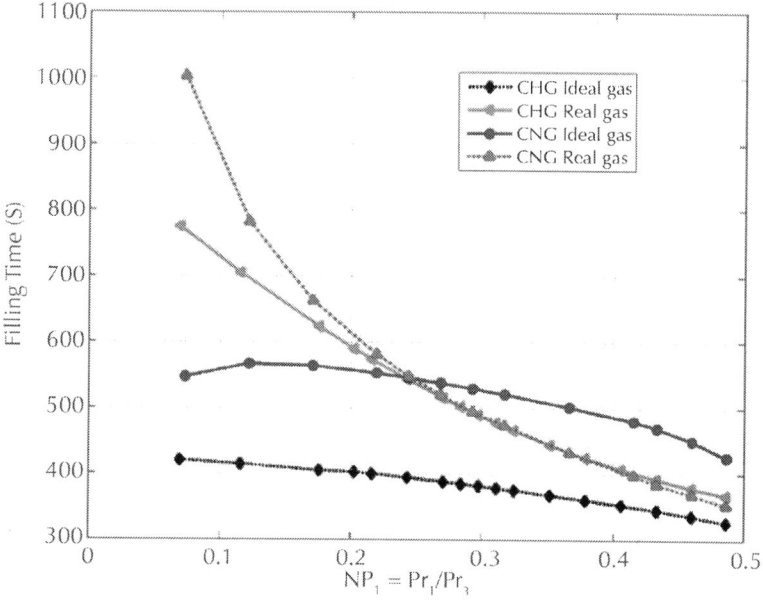

Figure 9: Effect of varying LPR pressure (NP1) on filling time (NP2 = 0.53) for the ideal and real gas models in the two stations (CNG and CHG).

Figure 10 shows the effects of MPR pressure on filling time when the LPR pressure is kept constant (NP1 = 0.275) for the real and ideal gas models in CNG and CHG stations. Considering the fact that reducing the filling time is a way to enhance filling station performance, a designer should seek a combination of an in which the filling time is minimized. For a constant value of NP1 = 0.275, there are specific values of for the models in which the filling time is maximized. So one can conclude for the real and ideal gas models in both stations (CNG and CHG) that the maximum values of the filling time are obtained as and respectively.

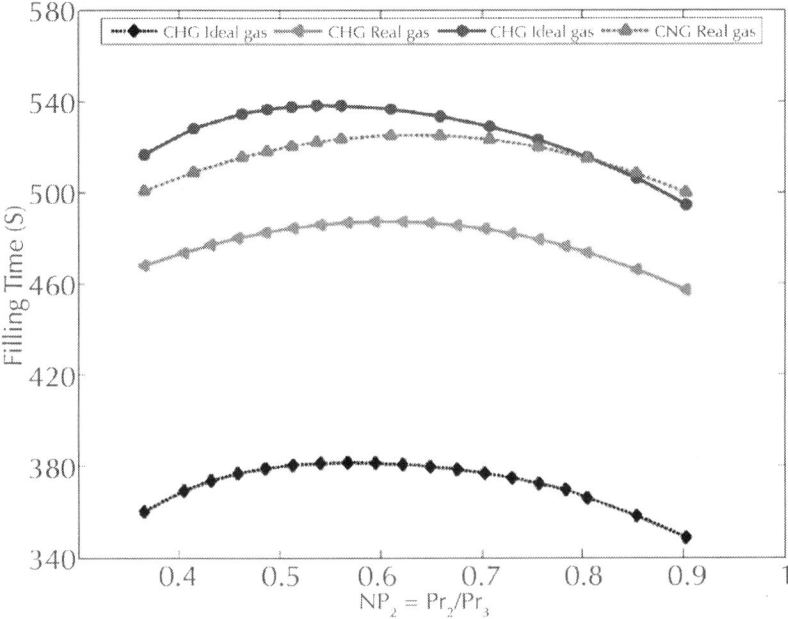

Figure 10: Effect of varying MPR pressure (NP2) on filling time (NP1 = 0.275) for the ideal and real gas models in the two stations (CNG and CHG).

It should also be noted that the filling time could also be reduced by appropriate sizing of the piping equipments (e.g., the orifice diameter).

As mentioned previously, entropy generation is associated with thermodynamic irreversibilities. Irreversibilities dissipate work into heat in the filling station. The available work is provided by the compressor, so one can conclude that, as entropy generation is decreased, available work destruction is decreased too.

Figure 11 shows the effects of LPR pressure on non-dimensional entropy generation when the MPR pressure is kept constant (NP2 = 0.53) for various gas models. For any gas model and station, the minimum values of non-dimensional entropy generation occur near NP1 = 0.22.

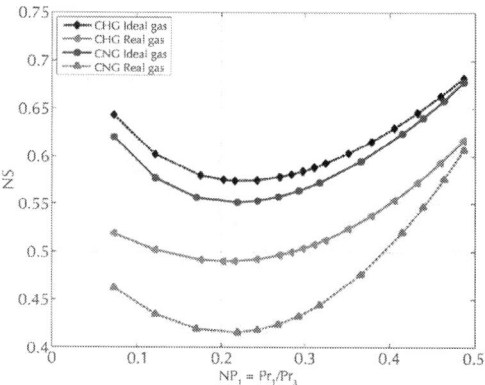

Figure 11: Effect of varying LPR pressure (NP1) on non-dimension entropy generation (NP2 = 0.53) for the ideal and real gas models in the two stations (CNG and CHG).

Figure 12 shows the effects of MPR pressure on non-dimensional entropy generation when the LPR pressure is kept constant (NP1 = 0.275). In both stations (CNG and CHG) the non-dimensional entropy generation for real gas is less than for the ideal gas model. Note from the figure that there is an optimum near NP2 = 0.6 where minimum entropy generation occurs.

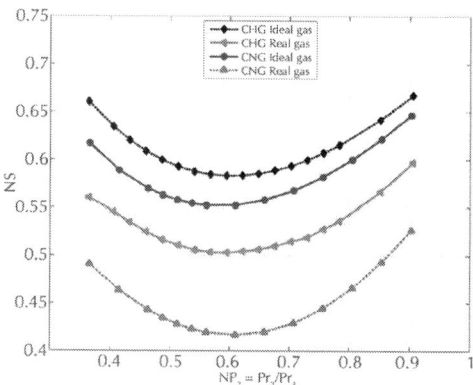

Figure 12: Effect of varying MPR pressure (NP2) on non-dimension entropy generation (NP1 = 0.275) for the ideal and real gas models in the two stations (CNG and CHG).

Considering Figures 9-10 and 11-12, it could be realized that, for each of the models in various stations, non-dimensional entropy generation and filling time profiles have opposite trends. So as entropy generation in the system decreases, the filling time increases. Because the filling time could be reduced by appropriate sizing of the piping equipment, one can conclude that an optimized thermodynamic point should be selected to enhance performance of a fuelling station.

CONCLUSIONS

The first and second laws of thermodynamics have been employed as theoretical tools to compare performance of cascade reservoir natural gas and hydrogen fuelling stations based on ideal and real gas models. A theoretical analysis has been developed to study and compare effects of reservoir temperature and pressure on fill ratio, filling time and entropy generation during the fast-fill process of on-board CNG and CHG cylinders.

It is found that, for both stations (CNG and CHG), as the reservoir temperature decreases, the fill ratio increases. The pressure within the filling station has no effect for ideal gas assumptions. The final in-cylinder temperature for CHG is much higher than for CNG, so for developing standards for CHG cylinders, this has to be considered.

For the real and ideal gas models in both stations (CNG and CHG), the maximum values of filling time are obtained when and, respectively. This is the case when the LPR pressure is kept constant (NP1 = 0.275).

Non-dimensional entropy generation and filling time profiles have opposite trends and, as entropy generation in the system decreases, the filling time increases. Since the filling time could be reduced by appropriate sizing of piping equipments, one could conclude that the optimized thermodynamic point should be selected for enhancing the performance of a fuelling station. For these filling stations (CNG and CHG), the optimized non-dimensional LPR and MPR pressures are found to be NP1 = 0.22 and NP2 = 0.6, respectively. In the present study, some simplifications have been employed such as the adiabatic assumption for the on-board cylinder and fixed conditions of the reservoirs, which will be studied in our future work.

REFERENCES

1.   Baur, S., Sheffield, J., Enke, D., First annual university students' hydrogen design contest: Hydrogen Fuelling Station, National Hydrogen Association, U.S. Department of Energy and Chevron Texaco (2004).

2.   Bejan, A., A study of entropy generation in funda-mental convective heat transfer. J. Heat Transfer, 101 718-725 (1979).

3.   Bejan, A., Entropy Generation Minimization. CRC, Boca Raton, NY (1996).

4.   Bejan, A., Second-law analysis in heat transfer and thermal design. Adv. Heat Transfer, 15, 1-58 (1982).

5.   Chan Kim, S., Hoon Lee, S., Bong Yoon, K., Thermal characteristics during hydrogen fuelling

6.   Deymi-Dashtebayaz, Farzaneh-Gord, M., Rahbari H. R., Studying transmission of fuel from storage bank to NGV cylinder in CNG fast filling station. Journal of the Brazilian Society of Mechanical Sciences and Engineering, Vol. XXXIV, pp. 429-435 (2013).

7.   Díaz, E., Cazurro, A., Ordóñez, S., Vega, A., Coca, J., Determination of solubility parameters and thermodynamic properties in hydrocarbon-solvent systems by gas chromatography. Braz. J. Chem. Eng., Vol. 24, No. 2 (2007).

8.   Dicken, C. J. B., Mérida, W., Measured effects of filling time and initial mass on the temperature distribution within a hydrogen cylinder during refuelling. Journal of Power Sources, 165, 324-336 (2007).

9. Farzaneh-Gord, M., Compressed natural gas single reservoir filling process. Gas International Engineering and Management, Vol. 48, No. 6, pp. 16-18, July/August (2008).

10. Farzaneh-Gord, M., Deymi-Dashtebayaz, M., Optimizing natural gas fuelling station reservoirs pressure based on ideal gas mode. Polish Journal of Chemical Technology, Vol. 15, No. 1, pp. 88-96 (2013).

11. Farzaneh-Gord, M., Deymi-Dashtebayaz, M., Rahbari, H. R., Niazmand, H., Effects of storage types and conditions on compressed hydrogen fuelling stations performance. International Journal of Hydrogen Energy, Vol. 37, pp. 3500-3509 (2012c).

12. Farzaneh-Gord, M., Deymi-Dashtebayaz, M., Rahbari, H. R., Optimizing compressed natural gas filling stations reservoir pressure based on thermody-namic analysis. Int. J. Exergy, 10, 299-320 (2012b).

13. Farzaneh-Gord, M., Deymi-Dashtebayaz, M., Rahbari, H. R., Studying effects of storage types on performance of CNG filling stations. Journal of Natural Gas Science and Engineering, Vol. 3, 334-340 (2011).

14. Farzaneh-Gord, M., Eftekhari, H., Hashemi, S., Magrebi, M., Dorafshan, M., The effect of initial conditions on filling process of CNG cylinders. The second International Conference on Modelling, Simulation, and Applied Optimization, Abu Dhabi, UAE, March 24-27 (2007).

15. Farzaneh-Gord, M., Hashemi, S. H., Farzaneh-Kord, A., Thermodynamics analysis of cascade reservoirs filling process of natural gas vehicle cylinders. World Applied Sciences Journal, 5, (2), 143-149 (2008).

16. Farzaneh-Gord, M., Rahbari, H. R., Deymi-Dashtebayaz, M., Effects of natural gas composi-tions on CNG fast filling process for buffer storage system. Oil & Gas Science and Technology, Rev. IFP Energies Nouvelles (2012a).

17. Kountz, K., Liss, W., Blazek, C., A New Natural Gas Dispenser Control System. Paper at 1998 Interna-tional Gas Research Conference, San Diego, November 3 (1998).

18. Kountz, K., Liss, W., Blazek, C., Automated Process and System for Dispensing Compressed Natural Gas. U. S. Patent 5,810,058, Sept. 22 (1998).

19. Kountz, K., Liss, W., Blazek, C., Method and Apparatus for Dispensing Compressed Natural Gas. U. S. Patent 5,752,552, May 19 (1998).

20. Kountz, K., Modelling the fast fill process in natural gas vehicle storage cylinders. American Chemical Society Paper at 207th National ACS Meeting, March (1994).

21. Kountz, Kenneth, J. and Blazek, Christopher. F., NGV Fuelling Station and Dispenser Control Systems, Report GRI-97/0398. Gas Research Institute, Chicago, Illinois, November (1997).

22. Liss, W. E., Richards, M. E., Kountz, K., Kriha, K., Modeling and testing of fast-fill control algo-rithms for hydrogen fuelling. National Hydrogen Association Meeting, March (2003).

23. Liss, W. E., Richards, M., Development of a Natural Gas to Hydrogen Fueling Station. Topical Report for U. S. DOE, GTI-02/0193, Sept. (2002).

24. Newhouse, N. L., Liss, W. E., Fast filling of NGV fuel containers. SAE Paper, 01-3739 (1999). Of Hydrogen Energy, 35, 6830-35 (2010). Process of type IV cylinder. International Journal

25. Shipley, E., Study of natural gas vehicles (NGV) during the fast fills process. Thesis for Master of Science, College of Engineering and Mineral Resources at West Virginia University (2002).

26. Sordi, A., da Silva, E. P., Neto, A. J. M., Lopes, D. G., Pinto, C. S., Araújo, P. D., Thermodynamic simulation of biomass gas steam reforming for a solid oxide fuel cell (SOFC) system. Braz. J. Chem. Eng., vol. 26, no. 4 (2009).

27. Sordi, A., Silva, E. P., Milanez, L. F., Lobkov, D, D., Souza, S. N. M., Hydrogen from biomass gas steam reforming for

low temperature fuel cell: Energy and exergy analysis. Braz. J. Chem. Eng., vol. 26 no.1 (2009).

28. Thomas, G., Goulding, J. and Munteam, C., Measurement, approval and verification of CNG dispensers. NWML KT11 Report, (2002).

29. Yang, J. C., A thermodynamic analysis of refuelling of a hydrogen tank. International Journal of Hydrogen Energy, Vol.34, pp6712-21 (2009).

30. Zheng, J., Ye, J., Yanga, J., Tang, P., Zhao, L., Kern, M., An optimized control method for a high utilization ratio and fast filling speed in hydrogen refuelling stations. International Journal of Hydrogen Energy, 35, 3011-17 (2010).

Chapter 7

Effect of Bed Deformation on Natural Gas Production from Hydrates

Mohamed Iqbal Pallipurath

Mechanical Engineering Department, TKM College of Engineering, Kollam, Kerala 691005, India

ABSTRACT

This work is based on modelling studies in an axisymmetric framework. The thermal stimulation of hydrated sediment is taken to occur by a centrally placed heat source. The model includes the hydrate dissociation and its effect on sediment bed deformation and resulting effect on gas production. A finite element package was customized to simulate the gas production from natural gas hydrate by considering the deformation of submarine bed. Three sediment models have been used to simulate gas production. The effect of sediment deformation on gas production by thermal stimulation is studied. Gas production rate is found to increase with an increase in the source temperature. Porosity of the sediment and saturation of the hydrate both have been found to significantly influence the rate of gas production.

INTRODUCTION

Energy demand is on the rise globally but the production rates of major fossil fuels are going down. Several analysts predict a drastic reduction in energy production due to diminishing reserve of fossil fuels. The major result from the global analysis is that world oil production peaked in 2006. Production has started to decline at a rate of several percentages per year. This necessitates a search for commercially viable and clean source of energy capable of meeting future energy demands. Natural gas hydrate (NGH) is one of the possible energy sources to meet these requirements. It is a highly condensed form of natural gas formed by capture of natural gas molecules in a cage of water molecules: each cubic meter of natural gas hydrate yields about 160 cubic meter of gas at STP.

A large amount of natural gas hydrate exists on our planet. Such deposits are found both on land (in the permafrost region), and offshore (in the submarine sediment). Over 230 gas hydrate deposits have been found globally. Gas hydrates have also been located in the coastal regions of India [1]. Needless to say, the vastness of gas hydrates has attracted global attention for its exploration and exploitation for future energy supply. It is predicted that utilization of even 17% to 20% of this resource could meet the energy demands for next 200 years [2].

Methods suggested for the production of natural gas from gas hydrate include depressurization, thermal stimulation, chemical inhibitor injection, and CO_2 sequestration. Among these, depressurization and thermal stimulation have been considered to be the most economical, though other methods are under investigation. The type of method depends on the reservoir characteristics. Due to less energy input for depressurization, this method has been studied more than the thermal stimulation. However, the efficacy of latter method needs to be studied in more detail. In recent times, during December 2001 to March 2002, field tests were conducted at a permafrost region located at Mallik gas hydrate site (Canada) in which both depressurization and thermal stimulation were conducted. Each of these methods has its relative merits and demerits, and for the development of commercial technology it is essential to have a good understanding of each of the proposed methods.

CURRENT SCENARIO

The study of any of the methods for gas production from natural gas hydrates needs not only a proper knowledge of the thermodynamics, kinetics, and heat and mass transfer effects, but also that of sediment response to any change in its morphology during hydrate dissociation. The presence of hydrate has a cementing effect on the sediment structure and hence any loss of hydrate tends to weaken the sediment. Sediment deformation has profound influence on the gas productivity. Moreover, a reduction in the sediment strength may destabilize sediment matrix to an extent that could cause serious damage to the production operation and uncontrolled release of methane to the environment.

The earlier modeling on gas production from gas hydrates considered an undeformed bed as the main objective was to study the effectiveness of a given method in the gas production. Some of these works considered the kinetics of hydrate dissociation (e.g., [3]), and others assumed instantaneous dissociation and hence equilibrium condition during hydrate dissociation (e.g., [4]). Also, the heat transfer effect was not considered in all the studies (e.g., [3] assumed isothermal operation). Analytical solutions were obtained by Selim and Sloan [4], Yousif et al. [3], Tsypkin [5], Ji et al. [6], and many others. However, such solutions were based on assumptions which were found not to be valid in reality. Numerical approach to the problem has been attempted to address the real conditions existing in the reservoir and other operational issues. Gas production from hydrate reservoir by the combination of warm water flooding and depressurization was proposed by Bai and Li [7] which can overcome the deficiency of single production method. Sun et al. [8] developed a one-dimensional model of hydrate depressurization in porous media.

Recently, a research group led by Professor Kimoto [9, 10] from Kyoto University, Japan, reported studies on gas production from gas hydrates considering bed deformation. They proposed a model based on chemothermomechanically coupled analysis which could predict the deformation of sediment; they did not report the gas production under this condition. Gas production from gas hydrate was also studied by some researchers by using some of the reservoir simulators such as CMG-STARS, Hydrate Res Sim, MH-21 HYDRES, STOMP-HYD, and TOUGH-HYDRATE.

Kinetics and thermodynamics of hydrate formation and dissociation dictate the choice of operating conditions and hence the gas production. Pioneering work on methane hydrate kinetics was done by Professor Bishnoi and his research team from University of Calgary (e.g., [11–13]). Other studies on methane hydrate kinetics were performed with different additives such as polymeric inhibitor [14], electrolyte solution [15], and promoter [16] A kinetic rate model was proposed by Kim et al. [13] for methane hydrate decomposition. Kinetics of hydrate formation of gases other than CH_4, such as C_2H_6, and natural Gas, was studied by Kaschiev and Firoozabadi [17].

The gas hydrate thermodynamics dictates the pressure-temperature relationship to predict the zone of hydrate stability. Also, the heat of dissociation of hydrate is obtained from such studies. Selim and Sloan [4] reported a thermodynamic relationship based on Antoine equation and also an equation to determine the heat of hydrate dissociation. Lu and Sultan (2008) summarized the studies on hydrate stability and proposed a correlation to relate pressure and temperature at hydrate equilibrium which gave a good match with the previous experimental data.

The effects of some polymers and surfactants on methane hydrate formation were investigated by Karaaslan and Parlaktuna (2004) in a high-pressure system.

Results of field test on gas production by thermal stimulation of hydrated sediments performed at Mallik 5L-38 gas hydrate production research well were reported by Hancock et al. [18]. No other field data on thermal stimulation is available.

A review of the reported studies in the literature shows that the following issues need to be investigated in more detail.

- The type of soil mechanical model that can best predict the production behaviour during thermal stimulation has not previously been investigated. The previous work available in the literature on the effect of soil deformation indicates that soil model may have a significant bearing on the gas production.
- Till now, experimental data are available only from one field test on thermal stimulation in permafrost region. It is necessary to conduct experiments to study this method under submarine conditions.

Description of the Model Used

Numerical modelling can be an effective tool that enables understanding mechanisms leading to wellbore instability in oceanic hydrate bearing sediment. To assess deformations caused by hydrate dissociation and the effect of these deformations on the gas generation, numerical techniques are essential. Thermal dissociation of hydrated sediment by a pumped hot fluid is modeled. A radial heat flow from the hot pipe is assumed. The coordinate system is cylindrical. Four components (soil, hydrate, gas (methane), and water) and three phases (hydrate, gas, and aqueous-phase) are considered in the simulator. The intrinsic kinetics of hydrate formation or dissociation is considered using the Kim-Bishnoi model. Mass transport and heat transfer involved in formation or dissociation of hydrates are included in the governing equations. The arrangement of heat source is shown in Figures 1 and 2.

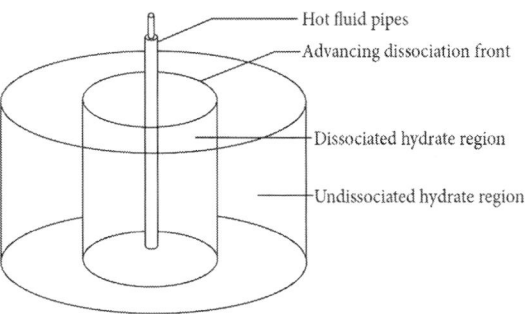

Figure 1: Schematic of a hydrate reservoir heated with pipes carrying hot fluid.

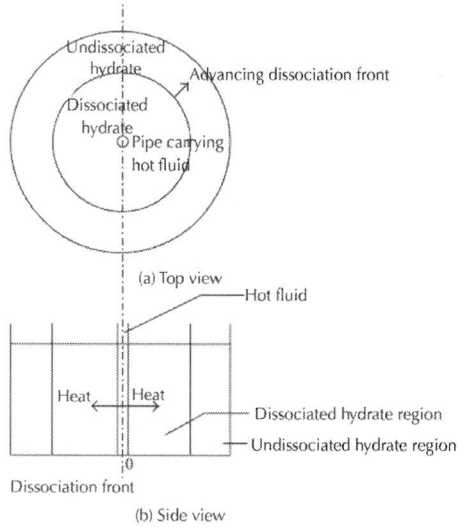

Figure 2: Schematic representation of hydrated sediment to study gas hydrate dissociation by thermal stimulation.

Factors to be evaluated are as follows.

- Changing the stresses and pore pressures.
- Impact of the selected constitutive model for pore pressures.
- Effect of heating the formation on thermodynamic stability of the hydrates.
- Deformation of the sediment as a result of dissociation of hydrate.
- Effect of this deformation on the gas production rates.
- Effect of different saturations of hydrate on gas production.
- Effect of source temperature on gas production.
- Effect of porosity on gas production.

Mechanisms to be considered are as follows.

- Kinetic rate, heat and mass transfer equilibrium, and fluid flow relations for gas hydrate dissociation/reformation with the change in pressure and temperature.
- Resultant changes in the mechanical and petrophysical properties of the sediments.
- Representative constitutive equation and yield criterion for the mechanical behavior of HBS of various hydrate concentrations.
- Different soil models and their responses to deformation.

The generation rates of gas and water are dictated by stoichiometry of the hydrate. The transition between gas, water, and hydrates can be represented as a chemical reaction

$$g\left(\frac{G}{A}\right) + n \cdot w\left(A\right) \Longleftrightarrow h\left(H\right), \tag{1}$$

where g is the gas component, existing as free gas (G) or dissolved in water (A), h is the hydrate component present only in the hydrate phase (H), and n is the hydration number.

Assumptions

- Gas hydrate bearing zone at a total depth of 3000 to 3300 meters (~700 m below sea floor and 2400 m below sea level) is considered the default range of study (Yun et al., 2010).
- We neglect the adsorption of any component by the rock phase; that is, the rock phase is inactive in mass transfer.
- Momentum (fluid flow) and heat transfer are axisymmetric.
- The water phase is incompressible.
- The gas follows the Peng Robinson equation of state.
- The porous medium (rock) is nondeformed.
- Gas can occur only in gaseous and hydrate states since CH_4 solubility under conditions of model is negligible.
- Water can occur only in liquid and hydrate states; that is, ice and water vapour formation are neglected since the sediment conditions preclude its formation.

Equivalent thermal conductivity of hydrated sediment is given by the equation

$$\lambda = (1 - \varphi)\lambda_s + \sum_{j=h,g,w} \phi S_j \lambda_j.$$

(2)

For a porosity of 0.47, let us consider two extreme cases of hydrate saturation, namely, 0.8 and 0.1. For hydrate saturation of 0.8 and water and gas saturations of 0.1 each, the equivalent thermal conductivity is $3.968\,\mathrm{W\,m^{-1}\,K^{-1}}$ whereas for a hydrate saturation of 0.1 and water and gas saturations of 0.45 each, the equivalent thermal conductivity is $3.868\,\mathrm{W\,m^{-1}\,K^{-1}}$ which is a difference of just 2.53%. This shows that an assumption of invariant equivalent thermal conductivity is valid.

Flow through Porous Media Applied to Hydrate Bearing Sediment

A porous medium is modelled in Abaqus/Standard by a conventional approach that considers the medium as a multiphase material and adopts an effective stress principle to describe its behaviour. The porous medium modelling provided considers the presence of two fluids in the medium. One is the "wetting liquid," which is assumed to be relatively (but not entirely) incompressible. Often the other is a gas, which is relatively compressible. An example of such a system is marine hydrated sediment containing sea water and gas. When the medium is partially saturated, both fluids exist at a point; when it is fully saturated, the voids are completely filled with the wetting liquid.

The porous medium is modelled by attaching the finite element mesh to the solid phase; fluid can flow through this mesh.

Coupled Flow and Heat Transfer through Porous Media

Optionally, heat transfer due to conduction in the soil skeleton and pore fluid, as well as convection in the pore fluid, can also be modeled. This capability represents an enhancement to the basic pore fluid flow capabilities discussed in the earlier paragraphs and requires the use of coupled temperature-pore pressure elements that have temperature as an additional degree of freedom in addition to the pore pressure and the displacement components. When you use the coupled temperature-pore pressure elements, Abaqus solves the heat transfer equation in addition to and in a fully coupled manner with the continuity equation and the mechanical equilibrium equations. Only linear brick, first-order axisymmetric, and second-order modified tetrahedrons are available for modeling coupled heat transfer with pore fluid flow and mechanical deformation. Coupled temperature-pore pressure elements are not supported in Abaqus/CAE.

Total and Excess Pore Fluid Pressure

The coupled pore fluid diffusion/stress analysis capability can provide solutions either in terms of total or "excess" pore fluid pressure. The excess pore fluid pressure at a point is the pore fluid pressure in excess of the hydrostatic pressure required to support the weight of pore fluid above the elevation of the material point. The difference between total and excess pore pressure is relevant only for cases in which gravitational loading is important. Total pore pressure solutions are provided when the gravity distributed load is used to define the gravity load on the model. Excess pore pressure solutions are provided in all other cases, for example, when gravity loading is defined with body force distributed loads.

Transient Analysis

In this transient coupled pore pressure/effective stress analysis, the backward difference operator is used to integrate the continuity equation and the heat transfer equation: this operator provides unconditional stability so that the only concern with respect to time integration is accuracy.

For fully saturated flow analyses in which heat transfer is also modelled, the contributions to the model's stiffness matrix arising from convective heat transfer due to pore fluid flow are unsymmetric.

Partially Saturated Flow

In gas hydrate sediment analysis, we shall be dealing with partially saturated flow. In partially saturated flow cases, the corresponding guideline for the minimum time increment is

$$\Delta t > \frac{\gamma_w n^0 \left(1 + \beta v_w\right)}{6 k_s k} \frac{ds}{du_w} (\Delta \ell)^2,$$

(3)

where s is the saturation, k_s is the permeability-saturation relationship, ds/du_w is the rate of change of saturation with respect to pore pressure, n^0 is the initial porosity of the material, Δt is the time increment, γ_w is the specific weight of the wetting liquid, k is the permeability of the soil, V_w is the magnitude of the velocity of the pore fluid, β is the velocity coefficient in Forchheimer's flow law ($\beta = 0$) in the case of Darcy flow, and $\Delta\ell$ is a typical element dimension.

Automatic Incrementation

Since automatic time incrementation is left to Abaqus, three tolerance parameters are chosen. The accuracy of the time integration of the flow continuity equations is governed by the maximum wetting liquid pore pressure change Δu_w^{max}, allowed in an increment. Abaqus/Standard restricts the time increments to ensure that this value is not exceeded at any node (except nodes with boundary conditions) during any increment in the analysis.

Since heat transfer is modelled, the accuracy of time integration is also governed by the maximum temperature change, $\Delta\theta_{max}$, allowed in an increment. Abaqus/Standard restricts the time increments to ensure that this value is not exceeded at any node (except nodes with boundary conditions) during any increment of the analysis.

Mechanical Constitutive Models

The constitutive library provided in Abaqus contains a range of linear and nonlinear material models for all of these categories of materials. In general, the library has been developed to provide those models that are most usually required for practical applications. There are several distinct models in the library, and for the more commonly encountered materials, several ways of modeling the material are provided, each suitable to a particular type of analysis application. But the library is far from comprehensive: the range of physical material behavior is far too broad for this ever to be possible. If there is no model in the library that is useful for a particular case, Abaqus/Standard contains a user subroutine UMAT. In these routines the user can code a material model (or call other routines that perform that task). This "user subroutine" capability proved to be a powerful resource for the modelling of gas hydrate bearing sediment. From a numerical viewpoint, the implementation of a constitutive model involves the integration of the state of the material at an integration point over a time increment during a nonlinear analysis (the implementation of constitutive models in Abaqus assumes that the material behavior is entirely defined by local effects, so each spatial integration point can be treated independently). Since Abaqus/Standard is most commonly used with implicit time integration, the implementation must also provide an accurate "material stiffness matrix" for use in forming the Jacobian of the nonlinear equilibrium equations.

FEM Formulation of Gas Production from Sediment Bed

The Abaqus package was customized with user subroutines and modified input files to solve the model equations. Some typical results obtained for submarine sediment are given. Figure 3 is a comparison of the three developed soil models for the base case using simulation of cumulative gas production for comparison. The three models of soil dynamical system give comparable results as shown in Figure 3.

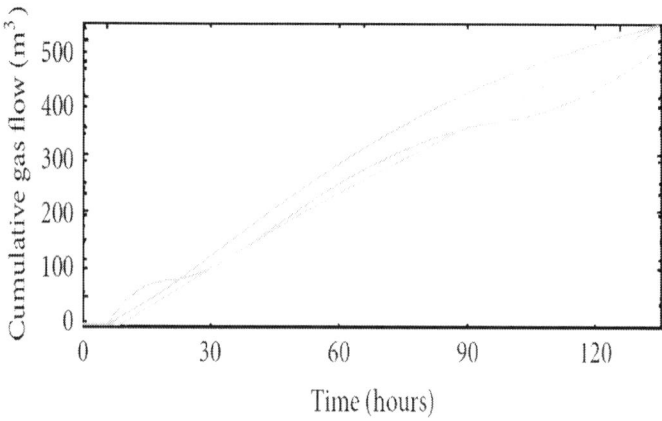

Figure 3: Three soil models compared in cumulative gas production rate.

In Figures 4 and 5, the temperature profiles for the viscoplastic and viscoelastic soil models are compared, and the figures indicate that, by around 60 hours, the temperature profile steadies and is unvarying with time thereafter. It may be inferred that, for a given radius of reservoir, there exists a constant period of time after which the temperature profile can be said to have achieved steady state.

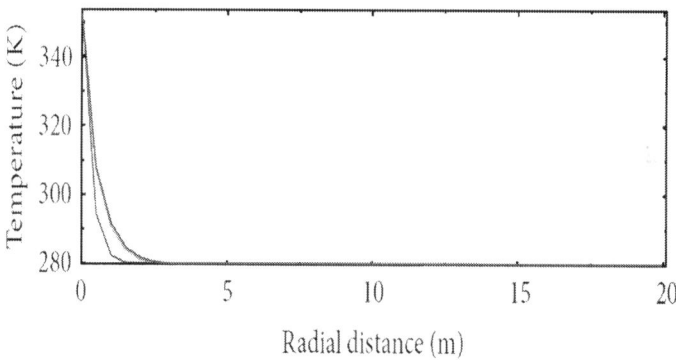

Figure 4: Temperature profile in sediment at various times using visco-plastic model.

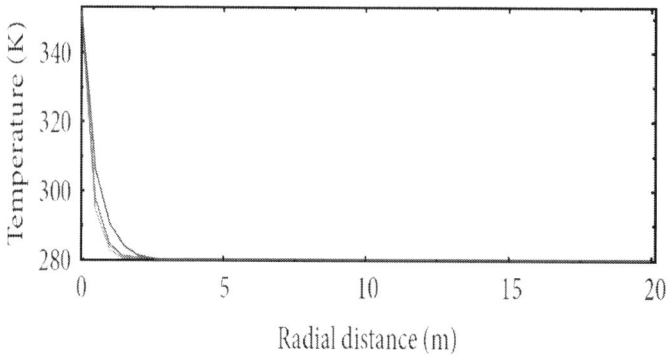

Figure 5: Temperature profile in sediment at various times using visco-elastic model.

It appears from the curves in Figures 4 and 5 that the steady state temperature distribution is not dependant on the soil model, provided the thermal properties of the sediment are more or less constant, as they are in these two soil models. But, as can be seen, the rates at which the steady states are achieved can vary marginally with the soil models used.

The effects of deformation on gas production are shown in Figures 6 and 7. Figure 6 shows cumulative gas production without deformation. When the plots are compared side by side, it appears that the amount of gas produced is reduced by more than 50% when we allow for deformation of the sediment. This could be explained by the reduced permeability of the sediment after dissociation and subsequent subsidence.

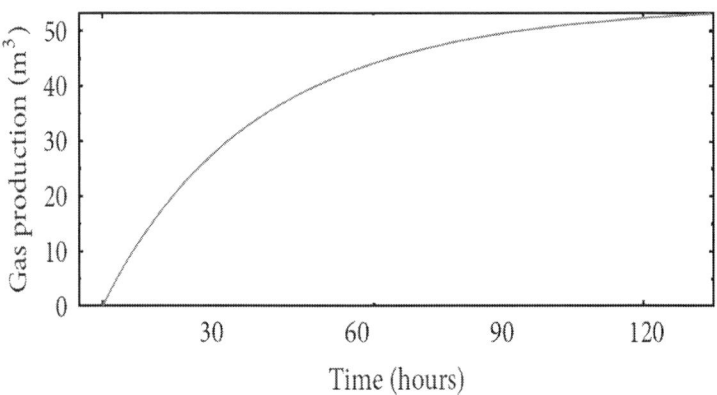

Figure 6: Cumulative gas production without deformation.

Figure 7 shows cumulative gas production with deformation of sediment bed using the viscoelastic soil model. The gaps between the sediment grains are reduced after subsidence, which could account for the drop in gas/water permeability, which in turn reduces the gas production by more than half.

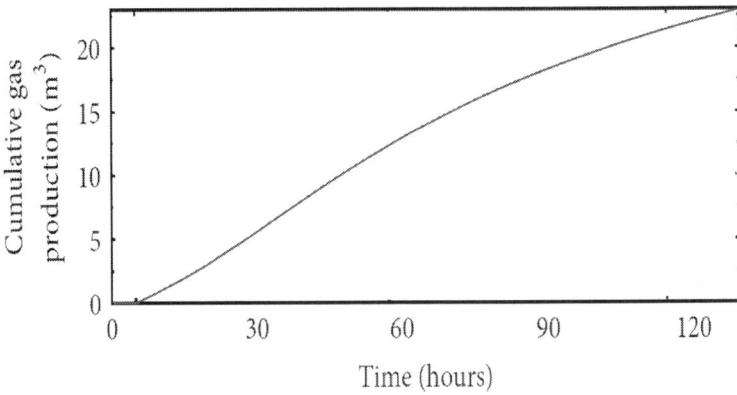

Figure 7: Cumulative gas production with deformation (viscoelastic soil model).

Parametric Study

A few representative results are shown of parametric studies conducted to gauge the impact of porosity and hydrate saturation upon gas production rate. The effect of sediment porosity upon gas production is graphically presented in Figure 8. Higher porosity seems to aid the production rate due to greater mobility for gas in higher porosity sediment (with higher permeability).

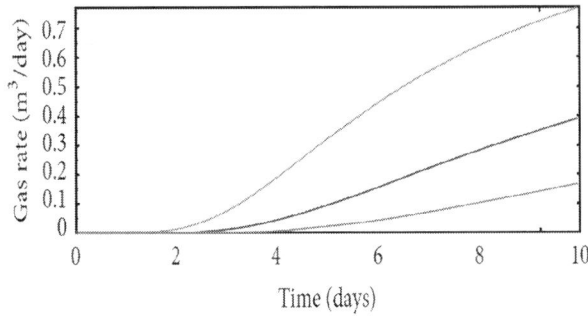

Figure 8: Effect of initial bed porosity on gas production rate.

Gas production rates at different initial hydrate saturations are shown in Figure 9. The production rates are dependent on initial hydrate saturation in a way that is expected, with higher saturations yielding greater amount of gas produced per day.

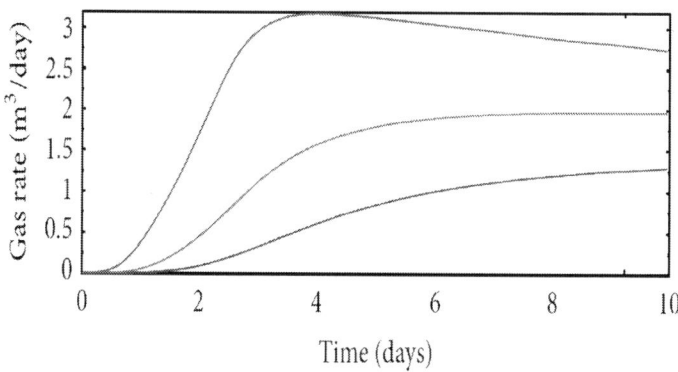

Figure 9: Effects of initial hydrate saturation on gas production rate.

For a constant temperature of the heat source, a higher saturation of hydrate is expected to correspond to larger volumes of produced CH_4 because of larger hydrate abundance. The substantial increase in the volume of the released gas when hydrate saturation increases from 0.5 to 0.8 as shown in Figure 9 confirms this expectation. Higher saturations of hydrate mean richer sediment and higher production rates. The effect of gas production rate on the heat source temperature is shown below in Figure 10.

The heat source temperature has a marked influence upon the rate of production of gas, with a factor of ~50% increase in production with 12.5% increase in temperature. But maintaining the heat source temperature at say 450 K is a costly proposition. A more economically sustainable temperature would be around 350–360 K.

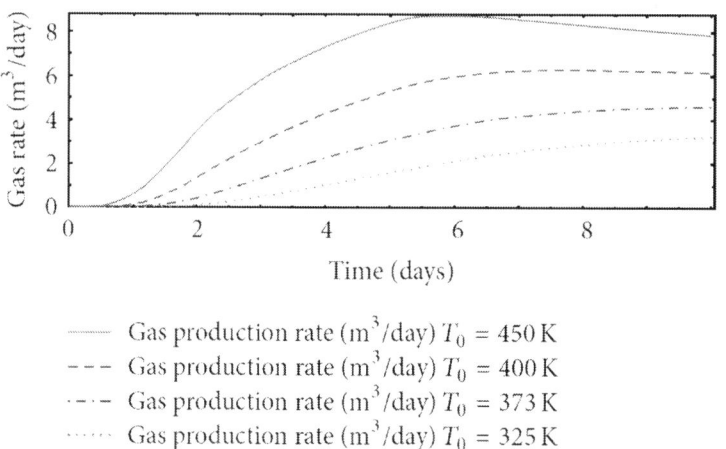

Figure 10: Effect of heat source temperature upon the gas production rate.

Validation of Model

The three soil models were validated with the well data from JAPEX/JNOC/GSC and others, Mallik 5L-38, gas hydrate production research well [18] as shown in Figures 11 and 12.

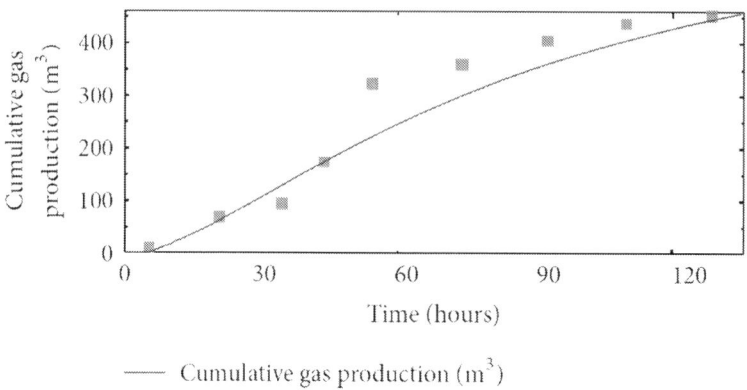

Figure 11: Poroelastic model of sediment validated with Mallik 5L-38 well data.

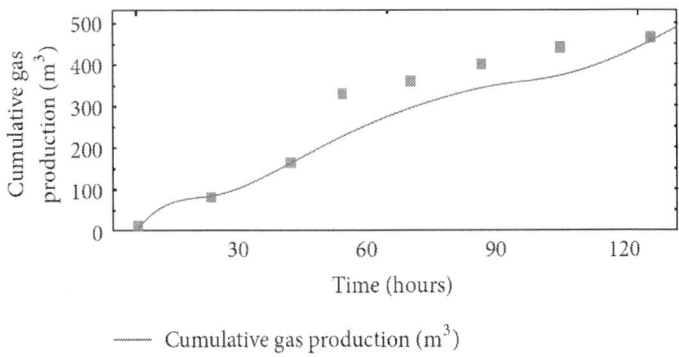

Figure 12: Viscoplastic model of sediment validated with Mallik 5L-38 well data.

FEM Profiles

Figures 13, 14, 15, 16, and 17 show the raw Abaqus screen output for parameters such as nodal temperature, temperature, and porosity across the width and depth of sediment.

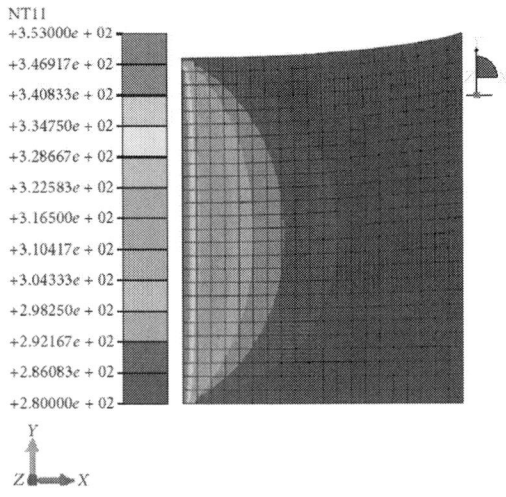

Figure 13: Nodal temperature profile of the sediment bed (with deforma-

tion). Radius of bed is 10 m, and depth is 13 m. Initial temperature was 280 K.

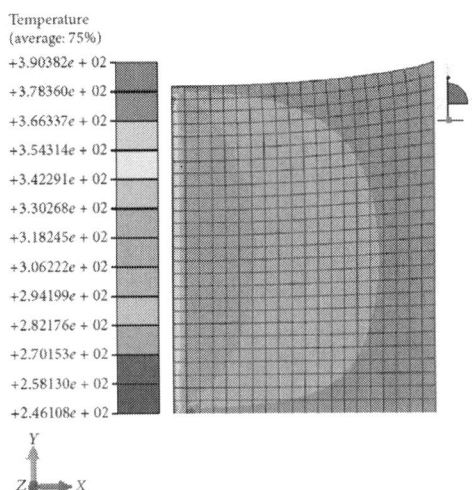

Figure 14: Temperature distribution after 50 days (with deformation).

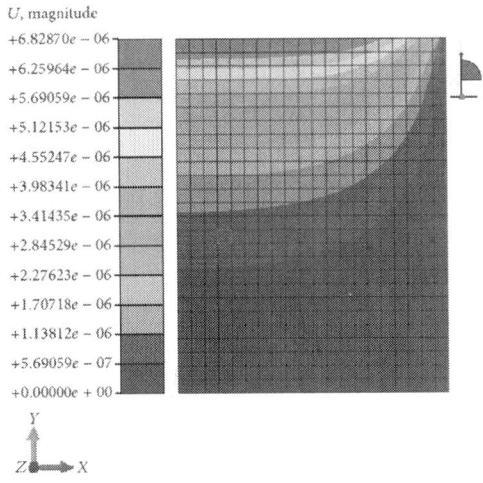

Figure 15: Magnitude of deformation (in m) after 6 hours.

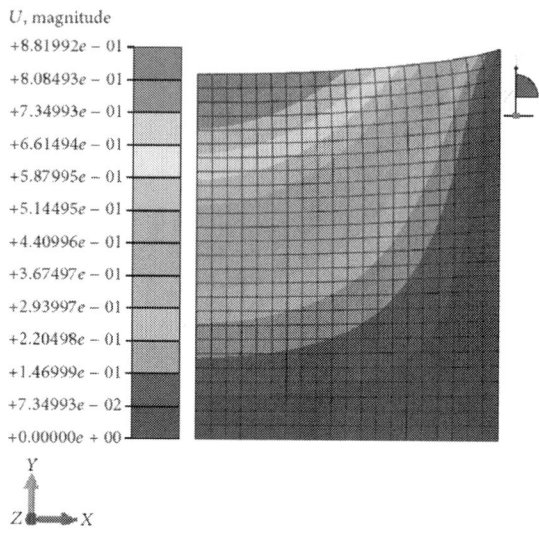

Figure 16: Magnitude of deformation after 50 days (m).

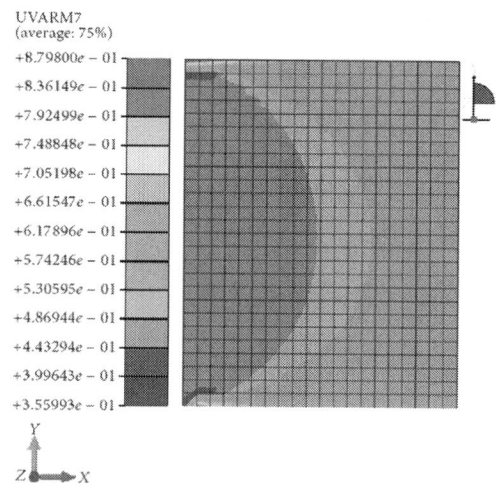

Figure 17: Porosity of sediment after 50 days.

CONCLUSIONS

A commercial package Abaqus was customized to simulate the gas production from natural gas hydrate by considering the deformation of submarine bed. The effect of sediment deformation on gas production by thermal stimulation is studied. The effect of three soil dynamic models, which are generally used for submarine sediments, on the prediction of the gas production has been studied. Gas production rate is found to increase with an increase in the source temperature. Porosity of the sediment and saturation of the hydrate both have been found to significantly influence the rate of gas production. The economics of the method of gas production from gas hydrates using thermal stimulation has been evaluated in the literature prior to this work. But such evaluations (Table 1) have always assumed that the gas production rate cannot be increased significantly by raising the temperature of the heating fluid. This study has found evidence to the contrary (Figure 10) which, though not conclusive, opens up new possibilities in the economic extraction of natural gas from ocean gas hydrate deposits worldwide in general and the KG and KK basins in India in particular.

Table 1: Economics of gas production from hydrates (modified from [19])

	Gas extraction methods		
	Thermal stimulation	Depressur-ization	Conventional gas production
Investment in millions of ₹*	254200	166000	157500
Annual cost in millions of ₹	160000	125500	100000
Total production (million m³/year)	1274.26	1557.43	1557.43
Production cost (₹/million m³)	6356.6	4025.9	3213.6
Break-even wellhead price (₹/million m³)	7945.8	5032.3	3972.9

*Calculated for currency rates at 2004 levels.

The higher overall annual average temperature of the regions where these basins are located do mitigate to some extent the energy overhead required to maintain a high temperature flow to the sediment bed.

The earlier techniques of hot fluid injection have many limitations as compared to the method studied in this work. In the earlier method, the heat losses to adjacent sediments are too big to permit economic extraction of gas from hydrate reservoirs by steam injection, even if the steam can be introduced at high rates (15 MW) into impenetrable hydrate reservoirs.

So also, low injection temperatures involve very big volumetric flow rates to carry useful amounts of heat into the reservoir. Injection of approximately $3600\,m^3$ per day of 66°C water is required if a heat flux of 15 MW is to be maintained. The limitations of disproportionate heat losses on the one hand and unrealistically high injection flow rates on the other hand will probably limit injection temperatures to between 66 and 120°C.

Similarly, unless the porosity is at least 0.15, the heat lost in raising the soil matrix temperature will render thermal stimulation (by injection) ineffective in producing useful quantities of gas. The relative importance of porosity in determining gas production is illustrated in Figure 8.

Scope for future work could be as follows.

- Consideration of inhomogeneity and anisotropy of the hydrate bed.
- Study of the interaction between the multiphase flow and the porous bed.
- Study of hybrid and alternate techniques for gas production.
- Further experimental validation of the above simulated results.

REFERENCES

1. "DGH Annual Activity Report 2010-2011," Hydrocarbon exploration and production activities.

2. Y. F. Makogon, "Natural gas hydrates: a promising source of energy," Journal of Natural Gas Science and Engineering, vol. 2, no. 1, pp. 49–59, 2010. · ·

3. M. H. Yousif, H. H. Abass, M. S. Selim, and E. D. Sloan, "Experimental and theoretical investigation of methane gas hydrate dissociation in porous media," in Proceedings of the SPE Annual Technical Conference & Exhibition, pp. 571–18320, October 1988.

4. M. S. Selim and E. D. Sloan, "Heat and mass transfer during the dissociation of hydrate in porous media," AIChE Journal, vol. 35, pp. 1049–1052, 1989.

5. G. G. Tsypkin, "Mathematical model for dissociation of gas hydrates coexisting with gas in strata,"Doklady Physics, vol. 46, no. 11, pp. 806–809, 2001. · ·

6. C. Ji, G. Ahmadi, and D. H. Smith, "Natural gas production from hydrate decomposition by depressurization," Chemical Engineering Science, vol. 56, no. 20, pp. 5801–5814, 2001. · ·

7. Y. Bai and Q. Li, "Simulation of gas production from hydrate reservoir by the combination of warm water flooding and depressurization," Science China Technological Sciences, vol. 53, no. 9, pp. 2469–2476, 2010. · ·

8. X. Sun, N. Nanchary, and K. K. Mohanty, "1-D modeling of hydrate depressurization in porous media,"Transport in Porous Media, vol. 58, no. 3, pp. 315–338, 2005. · ·

9. S. Kimoto, F. Oka, T. Fushita, and M. Fujiwaki, "A chemo-thermo-mechanically coupled numerical simulation of the subsurface ground deformations due to methane hydrate dissociation," Computers and Geotechnics, vol. 34, no. 4, pp. 216–228, 2007. · ·

10. S. Kimoto, F. Oka, and T. Fushita, "A chemo-thermo-mechanically coupled analysis of ground deformation induced by gas hydrate dissociation," International Journal of Mechanical Sciences, vol. 52, no. 2, pp. 365–376, 2010. · ·

11. A. Vysniauskas and P. R. Bishnoi, "A kinetic study of methane hydrate formation," Chemical Engineering Science, vol. 38, no. 7, pp. 1061–1072, 1983. ·

12. P. Englezos, N. Kalogerakis, P. D. Dholabhai, and P. R. Bishnoi, "Kinetics of formation of methane and ethane gas hydrates," Chemical Engineering Science, vol. 42, no. 11, pp. 2647–2658, 1987. ·

13. H. C. Kim, P. R. Bishnoi, R. A. Heidemann, and S. S. H. Rizvi, "Kinetics of methane hydrate decomposition," Chemical Engineering Science, vol. 42, no. 7, pp. 1645–1653, 1987. ·

14. J. S. Pic, J. M. Herri, and M. Cournil, "Experimental influence of kinetic inhibitors on methane hydrate particle size distribution during batch crystallization in water," Canadian Journal of Chemical Engineering, vol. 79, no. 3, pp. 374–383, 2001. ·

15. P. D. Dholabhai, N. Kalogerakis, and P. R. Bishnoi, "Kinetics of methane hydrate formation in aqueous electrolyte solutions," Canadian Journal of Chemical Engineering, vol. 71, no. 1, pp. 68–74, 1993. ·

16. N. Gnanendran and R. Amin, "Modelling hydrate formation kinetics of a hydrate promoter-water-natural gas system in a semi-batch spray reactor," Chemical Engineering Science, vol. 59, no. 18, pp. 3849–3863, 2004. · ·

17. D. Kashchiev and A. Firoozabadi, "Driving force for crystallization of gas hydrates," Journal of Crystal Growth, vol. 241, no. 1-2, pp. 220–230, 2002. · ·

18. S. H. Hancock, T. S. Collett, and S. R. Dallimore, "Overview of thermal-stimulation production test results for the JAPEX/JNOC/GSC et al. Mallik 5L-38 gas hydrate production research well," 2005.

19. B. S. Pierce and T. S. Collett, "Energy resource potential of natural gas hydrates," in Proceedings of the 5th Conference Exposition on Petroleum Geophysics, pp. 899–903, Hyderabad, India, 2004.

Citations

CHAPTER 1

Steve Larter, Haiping Huang, Jennifer Adams, Barry Bennett, Lloyd R. Snowdon, A practical biodegradation scale for use in reservoir geochemical studies of biodegraded oils, Organic Geochemistry, Volume 45, April 2012, Pages 66-76, ISSN 0146-6380, http://dx.doi.org/10.1016/j.orggeochem.2012.01.007.

CHAPTER 2

S. M. Peyghambarzadeh, A. Vatani, M. Jamialahmadi, and Experimental Study of Micro-particle fouling under Forced Convective Heat Transfer, http://dx.doi.org/10.1590/S0104-66322012000400004.

CHAPTER 3

Haytham M. Salem, Constantino Valero, Miguel Ángel Muñoz, María Gil-Rodríguez, Effect of integrated reservoir tillage for in-situ rainwater harvesting and other tillage practices on soil physical properties, Soil and Tillage Research, Volume 151, August 2015, Pages 50-60, ISSN 0167-1987, http://dx.doi.org/10.1016/j. still.2015.02.009.

CHAPTER 4

Vahid Tavakoli, Hossain Rahimpour-Bonab, Behrooz Esrafili-Dizaji, Diagenetic controlled reservoir quality of South Pars gas field, an integrated approach, Comptes Rendus Geoscience, Volume 343, Issue 1, January 2011, Pages 55-71, ISSN 1631-0713, http://dx.doi. org/10.1016/j.crte.2010.10.004.

CHAPTER 5

N. Spycher, L. Peiffer, E.L. Sonnenthal, G. Saldi, M.H. Reed, B.M. Kennedy, Integrated multicomponent solute geothermometry, Geothermics, Volume 51, July 2014, Pages 113-123, ISSN 0375-6505, http://dx.doi.org/10.1016/j.geothermics.2013.10.012.

CHAPTER 6

M. Farzaneh-Gord, M. Deymi-Dashtebayaz, and H. R. Rahbari, Effects of Gas Types and Models on Optimized Gas Fuelling Station Reservoir's Pressure, http://dx.doi.org/10.1590/S0104-66322013000200017.

CHAPTER 7

Mohamed Iqbal Pallipurath, "Effect of Bed Deformation on Natural Gas Production from Hydrates," Journal of Petroleum Engineering, vol. 2013, Article ID 942597, 9 pages, 2013. doi:10.1155/2013/942597.

Index